Quintessence

*Life's Essential Balance
between Stability, Novelty
and Fateful Encounters*

Michael Kuperstein

and

Richard G. Lanzara

Dedication

I dedicate this book, in fondest memory, to Dr. Michael Kuperstein (1954–2018) who strove mightily to understand our place in this marvelous universe. Michael's brilliant and fertile mind led to our tackling questions concerning life and evolution. This book is an attempt to introduce and embellish his original ideas, which I hope will do justice to his legacy. This book is also dedicated to our descendants.

—Richard G. Lanzara

Contents

CHAPTER 1

Beginnings

"Books are not absolutely dead things but do contain a potency of life in them to be as active as that soul was whose progeny they are; nay they do preserve as in a vial the purest efficacy and extraction of that living intellect that bred them."

—John Milton, *Areopagitica*

We're two scientists trying to understand life and evolution…

SOME BASIC QUESTIONS

- Can we understand the underlying scientific mechanisms of all life?
- Is there something fundamental about life that is structurally built into the universe?
- How does evolution place us on the cusp of all creation?

For a span of six years, Michael Kuperstein and I debated and discussed the science of life and evolution. This book is compiled from our discussions, e-mails, and notes. Our ideas might appear unusual or unique, but they've been well researched and discussed by us for years. This in no way insures their validity, but it represents our most direct and honest quest to arrive at a more substantial understanding of life and evolution. The discussion continues throughout this book as I have grappled with Michael's ideas with reference to my own. Hopefully, you, the reader, will recognize these tensions and become an engaged witness to this wonderful debate.

You are an amazing being! You represent the culmination of billions of years of evolution. Your chemical composition alone reveals that your chemical elements came from billions of years of evolution in the universe.[1] These are truly marvelous scientific facts!

On a fundamental level, we may wonder if life is somehow hard-wired into the structural design of the universe. This, naturally, leads to the question of whether life will be discovered on another planet or celestial body. It also raises the question of what we have yet to learn about the science of life.

Thinking in geological time is difficult for most of us. Having discovered fossils of sea creatures on mountain tops and the existence of deep-sea thermal vents, geology continues to find many amazing discoveries that are not only on Earth but also on other planets. If we consider the trillions upon trillions of stars, then the odds appear to be favorable that other life forms exist somewhere else in the universe.

Even today, we may be suddenly jolted by an exciting discovery of life miles beneath Earth![2] Although most of us don't spend much time considering the scientific ramifications of life, many of us do harbor an innate curiosity about life: where it might exist, what forms it may take, how it evolves. These are all questions that we may dabble in from time to time, leading us down many a rabbit hole into mind-bogglingly complex problems that seem unsolvable. You may already realize that we're at the very beginning of exploring and answering many of life's most baffling questions. This book is meant to engage

---

1   Neil DeGrasse Tyson and Donald Goldsmith, *Origins: Fourteen Billion Years of Cosmic Evolution* (New York: W. W. Norton & Company, 2004).

2   G. Borgonie et al., "Nematoda from the Terrestrial Deep Subsurface of South Africa," *Nature* 474 (June 2, 2011): 79–82, https://doi.org/10.1038/nature09974. See also Brandon Schmandt et al., "Dehydration Melting at the Top of the Lower Mantle," *Science* 344, no. 6189 (June 2014): 1265–1268, doi:10.1126/science.1253358; Oliver Plümper et al., "Subduction Zone Forearc Serpentinites as Incubators for Deep Microbial Life," *Proceedings of the National Academy of Sciences of the United States of America* 114, no. 17 (April 25, 2017): 4324–4329, https://doi.org/10.1073/pnas.1612147114; K. G. Lloyd et al., "Phylogenetically Novel Uncultured Microbial Cells Dominate Earth Microbiomes," *mSystems* 3:e00055-18 (American Society for Microbiology), https://doi.org/10.1128/mSystems.00055-18.

you on a journey of discovery, wonder, and awe as we explore and marvel at the wonders of life and evolution.

I have tried to capture the excitement Michael and I felt when exploring these fundamental questions, which we believe represent one of the great frontiers of science. You may find that this siren call of the unknown also draws you to examine and explore these paths. This book is for the adventurous reader who truly wants to explore how we became what we are. We try not to embrace the tyranny of fixed ideas while exploring the scientific edges of life and evolution. Ours is an intellectual journey; you don't need to leave home. As Einstein did with his *Gedanken* (thoughts) experiments, we reason and debate the path forward.

Michael and I have found science to be as beautiful and thrilling as a first love! It is often difficult to appreciate this unless one experiences it. This book encompasses Michael's and my explorations into the scientific areas that intersect life and evolution, which we have found both fascinating and vexing. Our natural curiosities drew us to explore many places where we might never otherwise have wandered. These innate curiosities are at the heart of this book and, we argue, are also a larger part of our lives, and all life, and evolution.

I knew Michael Kuperstein (1954–2018) for more than fifty years. Michael was the nerdy science type who went to MIT to get his PhD in neuroscience and later became an entrepreneur. Michael invented the world's first neural robot that learns from its own experience.[3] It wasn't until the last ten years of our friendship that we both realized our common interests in exploring the fascinating science behind life and evolution. It was during the last six years (2012–2018) that we began a regular back-and-forth correspondence that is distilled into this book. It was Michael's untimely death in 2018 that propelled me to write what we had discussed in our attempts to understand what brought us here at this very point in time within this marvelous universe.

During the years of our collaboration, Michael would often spontaneously call me with a new insight that needed further consideration, discussion, and debate. During these calls, his natural intelligence and insights into these

---

3   Michael Kuperstein, "Neural Network Model for Adaptive Hand-Eye Coordination for Single Postures," *Science* 239, no. 4845 (March 11, 1988):1308–1311, doi:10.1126/science.3344437.

difficult issues would prompt me to tell him to write it down. He'd replied that he would, but that it was all coming together in his head as a book he wanted to write. This is little consolation to me now, but Michael's son, Zack, provided me with Michael's notes and a large stack of research articles that Michael had kept. For this, I am grateful.

Michael and I had many lively debates and discussions about life and evolution. We exchanged hundreds of e-mails and had dozens of meetings. We didn't completely agree on all issues, but we did agree that there was a need for a book such as this. Writing this book posed multiple dilemmas. The massive amounts of scientific work that has been, and is being, constantly generated is too much for any one person to fully explore and comprehend. As this book evolved with a very large view, I realized that I had to focus its content on what I could write about. This includes some digressions that Michael and I had only briefly discussed. I expanded what we understood to be the role of sensors, membrane, and cytosolic memory, dormancy, and the emergence of information and complexity. This was because we had discussed these issues in passing with the expectation that we'd later return to what needed to be developed. Well, now that it is that "later," it is left for me to fill in the blanks. I learned that this wouldn't be the final word on these extremely interesting subjects, and that I was really writing only an outline of what we were trying to work out.

Most people believe that it is worthwhile to strive for and understand the latest scientific innovations and breakthroughs however long the journey and few the rewards. The public wants upbeat and positive stories when it comes to comprehending scientific studies, but the truth is that science is extremely complicated and often very messy. This book is where we reflect on our collective journey. Those who enjoy probing the unknowable may find this book useful as a framework that maps many new worlds in preparation for understanding how and why science and progress is messy by nature but awe inspiring nonetheless.[4]

---

4   Science is about skepticism, and no hypothetical scientific model is ever the definitive last word on a subject. While proposed models often appear good enough because they represent the reality of current scientific observations, they're only as good as the currently available data. At the early stages

As human beings, we are one with all life and the whole of creation up to this moment in time (see Appendix I A). We are all beings in this miraculous continuum that goes as far back as the beginning of life and even farther back to the beginning of the universe. We're brief witnesses to that amazing process and should be thankful that we're privileged to see it unfolding before us!

Stephen Hawking, the physicist who was confined to a wheelchair because he had amyotrophic lateral sclerosis (ALS, or Lou Gehrig's disease), is a heroic figure to me. In his show *Genius*, he demonstrated how we can find the origin of the big bang in the observable universe. Surprisingly, it doesn't matter where you are in the universe, you'll always find that the astronomical alignment of nearby celestial bodies points to a single place! This means that our entire observable universe appears to be the product of a single event that happened about fourteen billion years ago and that all of us, and our surrounding universe are the current progression of that single event. We are all centered on this expanding universe, which connects us to each other and the entire universe by the continual emergence of creation.

Hawking also had a TV episode on life in which he maintained that we're created from a random mixing of elements to form molecular machines that became the first life. He maintained that, then, evolution took over and directed our development as a response to environmental challenges that weeded out many life forms. He left us with the idea that we're the current end-products of that environmentally directed but rather random process. Hawking's basic understanding inspires us to seek a better and deeper understanding. Although much progress has been made in our knowledge of chemistry, biology and physics, we still have a long journey toward a complete understanding of life.

Studying the science of life should not detract from our awe and appreciation of it. As Richard Feynman stated:

Poets say science takes away from the beauty of the stars—mere globs of gas atoms. Nothing is "mere." I too can see the stars on a desert

---

of any proposed model, it's impossible to predict whether the model will survive to become scientific dogma, a somewhat depressing prospect as one would hope that science will remain ever exciting and fresh and never reduced to dogma.

night and feel them. But do I see less or more? The vastness of the heavens stretches my imagination; stuck on this carousel my little eye can catch one-million-year-old light. A vast pattern of which I am a part... What is the pattern, or the meaning, or the why? It does not do harm to the mystery to know a little about it. For far more marvelous is the truth than any artists of the past imagined![5]

Prepare for the path forward and take heart. Clear your mind for an intellectual dreamscape of fun and adventure!

Why Are We Here?

If you say, evolution, that may answer how we got here but not why. The answer to why we exist needs to include not only all of life's meaningful experiences but also the experiences of all of life that came before us, since we are the result of the evolution of prior species. The answer needs to include a grand continuity of life to capture the range of experiences from the simplest life forms to the most complex personal, social, and cultural experiences we have in our lives. From our very first development as human beings, our stories inherit basic phenomena that continuously compete and cooperate with each other throughout our lives. These phenomena are not only in effect during our lifetime but have been since biological life began almost four billion years ago. These biological phenomena have been a part of every species and every life form from the first bacteria to humans. Understanding what these are will have a profound influence on why we are here and the stories of our lives.

OUR INITIAL APPROACHES

When Earth's early chemistries (geochemistry) became life's early chemistries (biochemistry), life was born. It seems straightforward enough, but life has some rather unique properties that have eluded our scientific understanding.

---

5   Richard Feynman, "The Relation of Physics to Other Sciences," in *The Feynman Lectures on Physics*, vol. 1, lecture 3, section 3–4, "Astronomy," n.1 (New York: Basic Books, 1963).

What are these emergent properties and concepts that came from those processes that first formed life? Michael was acutely aware of these problems and resolved to find his way through the scientific difficulties surrounding life's beginnings. We believed that given our unique perspectives, Michael's PhD in neuroscience along with experience in computer science, and my PhD in biomedical sciences/pharmacology along with experience in physiology, biophysics, and toxicology, we might provide unique insights into the fascinating properties of life and evolution.

1994

In 1994, at Harvard University, fifteen researchers came together for a "Workshop on Self-Determination in Developing and Evolving Systems," which was given by Dr. Michael Kuperstein and Dr. Terry Deacon on January 6–9.[6]

---

6   Michael Kuperstein and Terry Deacon, "Workshop on Self-Determination in Developing and Evolving Systems," Harvard University, January 6–9, 1994. Researchers and attendees at this conference:
   1. Dr. Michael Kuperstein, Symbus Technology
   2. Dr. Terry Deacon, McLean Hospital, Harvard University
   3. Dr. Stephen J. Gould, Harvard University
   4. Dr. Dr. James Shapiro, University of Chicago
   5. Dr. Frank Ruddle, Yale University
   6. Dr. William Wimsatt, University of Chicago
   7. Dr. William Calvin, University of Washington
   8. Dr. Jason Brown, NYU Medical Center, Department of Neurology
   9. Dr. Peter Corning, Institute for the Study of Complex Systems
   10. Dr. Richard Ryan, University of Rochester
   11. Dr. Edward Deci, University of Rochester
   12. Dr. Thomas Ray, ATR HIP Japan
   13. Dr. David Ackley, Bellcore
   14. Dr. Jay Mittenthal, University of Illinois
   15. Dr. Domenico Parisi, CNR Institute of Psychology, Rome, Italy

Additional Attendees
Dr. Terry Allard, ONR program manager
Harry Stanton, A Bradford Book, MIT Press
Michael Waldrop, *Science*
Roger Lewin, *American Scientist* and *New Scientist*
John Rennie, *Scientific American*
George Johnson, *New York Times*

These researchers brought their collective expertise from biology, neuroscience, developmental psychology, and computational modeling. A glance at the list of attendees shows quite an impressive number of distinguished people. Interestingly, the most important consensus to emerge at the workshop was "the study of the emergence of Self across life forms has enormous potential for understanding phenomena in the life sciences and solving problems in computation."

In 1994, I also published an article demonstrating that a very common physiological law, Weber's law, also known as the Weber-Fechner law, could be derived by applying the physical laws of a simple balance to the measurement of the difference of weights.[7] This provided an important link between the underlying physical-chemical world and the world of our sensory perceptions. Since all our senses depend on molecules (receptors) that detect and respond to environmental changes, this finding also addressed a fundamental aspect that all life has in common: life must sense its environment if it is to survive, adapt, and evolve. This is also true for the earliest life, which had to sense whether the environment was helping or hindering its survival.

THE PROBLEM OF A GOOD STORY

We all like to read a good story with an orderly and logical progression of facts that coalesce into a satisfying whole. Our brains may even crave these completely satisfying stories. Yet if we were to go back to when an author first decided upon a story and witnessed the subsequent writes and rewrites until it finally became a publication, we'd be surprised at the number of alterations that eventually became the finished product. The same is true for this book and for other books on life and evolution. Sometimes these books present such a tidy, compelling story that sums up the whole business of something, such as "natural selection," that we buy it literally and figuratively.

These nice stories sound so pleasing to our ears but contain huge gaps in the most important scientific details and ideas. This book is an attempt

---

7   R. G. Lanzara, "Weber's Law Modeled by the Mathematical Description of a Beam Balance," *Mathematical Biosciences* 122, no. 1 (July 1994): 89–94, http://cogprints.org/4094/.

at preparing you, the reader, for potential discoveries. Because these topics are enormously difficult to understand, we'll only be skimming the surface of these vast lakes of knowledge, but that is perhaps the most that this book can do. Let's begin the process of placing these phenomena within the context of the beauty and magic of life so that we can also better appreciate the scientific thought that's required for us to begin to understand.

By nature, we're not often brutally honest with ourselves. Even very accomplished scientists are not always honest with themselves. It is human nature to make up a nice story to describe what we know little or nothing about. This applies more so to life and evolution. We might see a genetic change that appears to correlate with the enhanced survival of an organism, but what do we really know? Not much. We don't know how the change in the gene affects all the other genes and the subsequent metabolic reactions within an organism. Or maybe the organism survived because it was just lucky! There are enormous complications that we don't see.

Problems with Our Mental Constructs of Hereditary and Inheritance

Where did your oldest relatives come from, and their relatives, and so on? As you delve far back into your ancestry, you realize that there must have been some relatives that were not quite human. Before that, there must have been some relatives that were not mammalian. Before that, you probably had relatives that emerged from the sea, and long before that, some original cell represented the origin of life on Earth.

We can trace all our relatives back to the juncture when humans first emerged from their ancestors Thinking so far back in time, we're often confronted with a paradox which is true for other life forms as well. It goes like this: How could a new being come into existence and find the means to reproduce to create other new beings? For those creatures requiring sexual reproduction, there must be a second new being as a partner. This thinking has created quite a bit of trouble for our understanding of evolution. For example, we have Lucy, who was supposed to represent mankind's first ancestor, and Luca (last universal common ancestor) for all cellular life. Who or what were their counterparts and what other forms of life, or nonlife, could have supported their survival? These constriction points in our thinking have prevented us from making more progress. How should we think about these points of constriction? How might any unique being come into existence and stay in existence?

Let's make an analogy with the computers we have today. (One could just as well use anything that we recognize as a separate "thing," such as autos, bikes, clothes, etc.) How did the first computer arise? When we try to understand why at a certain point in time the computer was invented, we must understand the environmental circumstances that led to the development of a computer. We recognize that today's computers couldn't have been invented until the invention of electric current, among other factors. This level of understanding raises the bar significantly. For new life to form, many divergent lines branch and rejoin in large numbers of possible combinations that are, essentially, exploring the possible living spaces (niches or potentially habitable spaces) available. It isn't only one new entity that forms *de novo* but a myriad of slightly new entities that continue interacting with each other until a new line can develop and branch off the old lines of the tree of evolution. In our attempt to see the underlying patterns of evolution, we ignore the inherent messiness because it becomes too complex for our minds to grasp. We don't have a scientifically ordered or neat method to conceptualize these processes, but the messiness is the most important part.

We're just beginning to discover some of the myriads of possible chemistries that might have existed on early Earth. When we think about the origin of life, either deep within Earth, or at some deep-sea hydrothermal vent, or a warm little pond, we've isolated the ingredients and created a singularity in our thinking that limits our research. When the Wright brothers invented the first airplane that could fly effectively, they were among many others attempting to fly. There were also others who were "flying" by hot-air ballooning or gliding. The point is that the time and environments were ripe for experimentations into flying with machines. The essential tools also had to be invented and available to those who wanted to build one of these flying machines. This had to all come together as an ecologically evolving system for the invention of the first airplane. Once it was invented, information spread so that others could easily copy it and produce other forms.

So many processes of change and alteration along so many disparate lines of descent over such a long time suggests that our simplifications don't provide the full picture. Even something as straightforward a system as air travel becomes immensely more complicated when you go back and recreate all the myriad number of inventions and parts that combined for it to exist. How much more complex is the evolution of life!

This book presents new concepts of life and evolution as it attempts to define what shapes our lives and all life on Earth.

KEY IDEAS DISCUSSED WITHIN THIS BOOK

1. The idea that life is a compilation of five essential concepts (see Appendix I B). These concepts are:
   - Self
   - Essential Tensions
   - Growth and Energy Flow
   - Stability versus Novelty
   - Fateful Encounters

   The interruption of any one or more of the five concepts may either force the Self into a dormant state or destroy it.
2. The idea that all life has emerged from chemistries that balance Stability (homeostasis) with the search for Novelty, which is inherent within and vitally important to life.
3. Life internalizes Novelty as new internal chemistries (Quintessence).
4. The idea that two-state chemical systems act as biological sensors by a net shift in their equilibria (see Appendix III C).
5. The formation of membranes has multiple important functions:
   - separation of the Self from the environment
   - structural and chemical creation of the Essential Tensions (see Appendix III)
   - inheritance of past environments (embedded in the structure of the membrane)

- positioning of sensors (receptors) that sample the external environment
6. The idea that inheritance is an emergent phenomenon in which cytoplasmic transfer and membrane biogenesis play important roles.

CHAPTER DESCRIPTIONS

Chapter 1, "Beginnings," concerns the historical background of this book.

Chapter 2, "Evolutionary Dreams," discusses previous attempts to define life.

Chapter 3, "Essential Tensions," outlines life's necessary chemistries and the critical importance of sensory molecules.

Chapter 4, "Quintessence," reveals Michael's broader view of life and evolution.

Chapter 5, "Stability Versus Novelty," describes how life balances the search for novelty with stability.

Chapter 6, "Growth and Energy Flow," argues that life taps energy flows to grow.

Chapter 7, "Emergence," discusses the mysteries of emergence in nine areas:

1. Emergence of chemistry
2. Emergence of metabolism
3. Emergence of Self
4. Emergence of life
5. Emergence of information
6. Emergence of heredity
7. Emergence of evolution
8. Emergence of species
9. Emergence of complexity

Chapter 8, "Fateful Encounters," holds that Fateful Encounters have shaped life.

Chapter 9, "The Continuum," argues that life's emerging chemistries continue their search for Novelty and the subsequent attempts to embrace and incorporate it (Quintessence) into life have created multiple new paths that have emerged and will continue to emerge with surprising results.

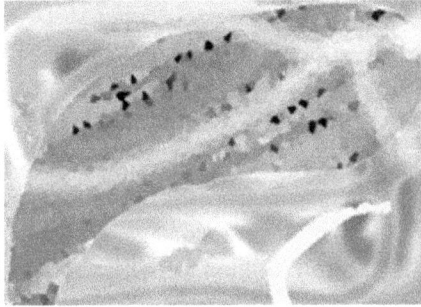

CHAPTER 2

Evolutionary Dreams

"Life is nothing but an electron looking for a place to rest."

—*Albert Szent-Györgyi*

The last sentence in Charles Darwin's *On the Origin of Species by Means of Natural Selection* reads as follows:

> There is grandeur in this view of life, with its several powers, having been originally breathed into a few forms or into one; and that, whilst this planet has gone cycling on according to the fixed law of gravity, from so simple a beginning endless forms most beautiful and most wonderful have been, and are being, evolved.

It is difficult to follow Darwin's eloquence; many have tried with varying degrees of success. Why are we trying? We believe that there's something we need to add. A very large amount of scientific work reveals a more detailed understanding of evolution and the laws of genetics (modern synthesis and other extensions of Darwinian theory that include extended evolutionary synthesis[1] and evolutionary developmental biology[2]).

---

1 *Wikipedia*, s.v. "extended evolutionary synthesis," https://en.wikipedia.org/wiki/Extended_evolutionary_synthesis
2 *Wikipedia*, s.v. "evolutionary developmental biology," https://en.wikipedia.org/wiki/Evolutionary_developmental_biology

These are all important contributions, but there still appear to be some missing elements that would be the next foothold for new progress. We believe that finding a basic definition of life would be that advance. After all, evolution is primarily about life, and perhaps it wasn't "from so simple a beginning," as Darwin states.

Even the word, *evolution* has itself emerged and evolved. How can we begin to understand this vast array of data from so many areas of biology, chemistry and physics? This is extremely difficult because one should know the science and terminology from these fields plus the science and terminologies from the field of evolutionary biology. Much as the blind men describing the elephant needed to keep their perspectives in check, we also need to do this to see our goal. Our minds haven't been programmed to understand and filter this vast array of data. With computers and artificial intelligence (AI), we might receive some assistance in the future, but we're not there yet.

Even our best computers today need a directed goal. What will they filter and search for in this vast array of data? Can we ask our computers what life is? Perhaps the best that our computers can do is to explain life as highly organized material entities composed of molecules that demonstrate the biochemical properties of life. To the question of what life is, the Google online dictionary answers:

1. the condition that distinguishes animals and plants from inorganic matter, including the capacity for growth, reproduction, functional activity, and continual change preceding death.
     synonyms: existence, being, living, animation

2. the existence of an individual human being or animal.
     synonyms: person, human being, individual, soul

These aren't scientifically satisfactory answers. The following definition might be a bit more satisfying: life is made up of the four qualities of metabolism, cell enclosures, growth, and reproduction that evolves. However, these words just skim over the truly interesting and more mysterious areas.

The view that defining life is a futile endeavor is common among biologists, most of whom do not attempt to undertake what appears to be a foolish task. This deferral may be surprising considering that science should provide clear and logically consistent definitions of all concepts employed. On the other hand, from the scientific point of view, the general concern should be to contribute as clear and universal an understanding of life as possible.

In the millennium year 2000, the science of biology celebrated its second centennial. Biology consists of two kinds of basic research, one dealing with biochemistry and physiology, and the other dealing with genetics and evolutionary theory. Modern biology texts often explain life much as our computers might attempt to answer that question. Below is *Wikipedia's* definition:

> Life is a characteristic that distinguishes physical entities that do have biological processes, such as signaling and self-sustaining processes, from those that do not, either because such functions have ceased, or because they never had such functions and are classified as inanimate.[3]

Reconciling this with the science of chemistry remains a major barrier to a better scientific understanding of life and evolution.

It was Louis Pasteur in the mid-nineteenth century who demonstrated that there was no such thing as spontaneous generation. However, many scientists feel that there is a link between early Earth's geochemical world and the later world of early life. Perhaps our imperfect understanding of early geochemistry coupled with the vast time spans and many extreme energies present on early Earth prevent science's current attempts to create life from a mixture of essential chemical elements. Given the myriad of possible chemistries, this may be too large a chasm to cross experimentally. Yet many scientists believe that if we could describe life, then recreating the necessary chemical processes may be

---

3 *Wikipedia*, s.v. "life," https://en.wikipedia.org/wiki/Life. Life refers to physical entities that have biological processes, such as signaling and self-sustaining processes, from those that do not, either because such functions have ceased, or because they never had such functions and are classified as inanimate. See also Maynard J. Smith, *The Problems of Biology* (Oxford: Oxford University Press, 1986), and Claus Emmeche, "Defining Life as a Semiotic Phenomenon," *Cybernetics and Human Knowing* 5, no. 1 (January 1, 1998): 3–17.

easier. Even supposing we could create a new chemical life form, would we be able to recognize it? What's chemical evolution? What would it have to do for us to recognize that it is alive? Would we be able to keep it alive? What would be its food or energy source? Would it survive away from its energy source? These are many of the same questions we need to ask about the origin of life on Earth.

What would we measure to determine whether our chemical experiments had become living? In our multiple efforts to create artificial protocells that might be alive, we may have already succeeded, but perhaps we were unable to detect that such cells were, indeed, living. It is more than an academic question. How can we know we have succeeded? What if the time frame for cell growth and division takes much longer than we are prepared to observe? What if it takes years or decades? Our first dreamscape is imagining life. What are the defining characteristics of life? Is it an enzyme in search of an electron or something much more?

A definition of life should describe the essential concepts (see below and Appendix 1 B) that are necessary for life to exist and evolve. These concepts are important both to ground us in the enormous task forward and to give us a scaffold on which we may construct a more natural and scientific way to advance our understanding. Whether or not one subscribes to another philosophy of science or thought, one should recognize the importance of a clear understanding of the distinctive features of living biological systems.

FIVE ESSENTIAL CONCEPTS DEFINING LIFE

We propose the following set of five essential concepts to describe life, which will be described in more detail throughout this book:

1. *Self versus non-Self.* By defining *Self,* we set it in opposition to non-Self. Self and non-Self are separated by a barrier, such as a semipermeable membrane. This definition of Self also requires the presence of the other four concepts listed below.

2. *Essential Tensions versus non-Essential Tensions.* These are sets of chemical systems that are coupled together in ways that prevent either one of them from realizing their normal chemical equilibrium. These

poised chemical imbalances are essential for all life, are sources of energy for metabolism, and include sensory systems.

3. *Growth and Energy Flow versus stasis (dormancy) and no Energy Flow.* This pertains to the uses and storage of extra molecules and/or energy and how the proteins and enzymes of various metabolic cycles control and direct energy into molecules that store energy or create new molecules such as fats, proteins and enzymes. Overall Growth requires various stages that eventually include the reproduction of the Self.

4. *Stability versus Novelty.* There is a dynamic balance between the stability-enforcing molecules of the Self and the novel expansion of potential physical and chemical spaces, such as for new energy sources. The potential chemical space of the Self may be expanded by novel encounters with new exogenous molecules or by the creation of new proteins/molecules from existing versions. This allows for many important modifications, including the chemical modifications of existing molecules by various secondary processes within cells. The interplay of Stability with Novelty produced mechanisms for the emergence of inheritance.

5. *Fateful Encounters versus no Fateful Encounters.* Fateful Encounters allow opportunities for new chemistries to link with an organism's existing metabolism and biochemistry. Often, these changes may be detrimental to the Self, but some may prove beneficial. Fateful Encounters can, literally, enrich the biochemistries of life and potentially lead to expansions by increasing physical, chemical, and biological encounters. They may also lead to more extreme encounters, such as mergers (symbiosis: a change through interdependence or union), or competition (e.g., eating or being eaten). These encounters depend on the type and number of environments the Self occupies.

Michael was an astute and brilliant scientist. He wanted to find the scientific bedrock upon which we could base a true scientific understanding of life and evolution. He thought we needed an approach with a set of basic concepts that could drive discovery forward. The recognition and search for these higher-level concepts excited and propelled our research forward. Michael's genius lay in his development of a system of specific concepts necessary for anything to be considered living, with the potential to evolve. Each of Michael's suggested, essential concepts are related to specific chemical and biological components. Taken together, they may assist us in exploring new and exciting paths to understanding life and evolution.

In the larger picture of evolution, we don't directly address the artificial divisions between Darwin and Lamarck. If one reads what they wrote, then one understands that later interpretations have given the false sense of an irreconcilable difference between them. We imagine that Darwin and Lamarck could have been friends who discussed their observations of nature without rancor. After Darwin made his careful observations, he concluded that a new species can originate from previous species by a series of changes over time. Lamarck observed that organisms can eventually change, based on their responses to environmental conditions. Neither Darwin nor Lamarck had the knowledge of the biochemistries of life that we have today, though we're only at the beginning of understanding how these biochemistries sense, respond, and adapt to environmental changes. It should be no surprise that environmental conditions can change life's biochemistries by the inherent feedback mechanisms embedded within all life. Today we know that these biochemistries are linked to the heritable, genetic chemistries of all life in many interesting and surprising ways.

The essential conditions for life to exist did not appear out of nowhere. They are themselves inherited from the physics and chemistry of early Earth: chemical gradients formed by the separation of different chemical and physical processes, and gradients formed from differences in temperature, chemistry, acidity, ionic concentrations, and pressure, across boundaries. These gradients drove precipitation of chemicals as well as ionic chemical flows. Some of these chemical precipitates formed compounds from strings and sheets of chemicals that self-assembled to create protocells, which were nonliving organic chemicals encased in membranes and resembling today's cells.[4] In addition,

---

4 William Martin and Michael J. Russell, "On the Origin of Biochemistry at an Alkaline Hydrothermal Vent," *Philosophical Transactions of the Royal Society B: Biological Sciences* 362, no. 1486 (November 3, 2006):1887–1926, doi:10.1098/rstb.2006.1881.

ionic gradients created flows of ions that were embedded within a protocell's membranes and became sources of energy production.[5]

These Energy Flows eventually evolved into the metabolism of digestion and breathing and fueled the first self-regulating metabolism that converted nourishment into cellular structures that maintained the protocell's stability. To survive in a changing environment, these early protocells were required to respond and adapt to novel aspects of their environment. The chemical gradients enabled the first metabolism, sensors, and eventually, as life began, inheritable forms. Thus arose the phenomena of seeking Novelty and maintaining Stability, fueled by the energy generated by chemical gradients.

We need to understand these essential properties of life in order to understand such phenomena at work in all cells. All life forms are composed of cells. Each cell is enclosed by a membrane distinguishing it from its environment and other cells. Each cell draws nourishment and energy from its environment, which feeds the cell's metabolism self-regulating its existence. In addition, each cell needs to be reactive to both its changing environment and its internal state. All life needs a self-regulating metabolism that maintains the cell's internal Stability. In doing so, the cell may orient toward or away from a changing environment to capture nutrients or escape dangers. These two choices appear to be in competition for the cell's energy since they drive toward very different results, one to maintain Stability and the other to engage in novel interactions with unknown future benefits.

How much energy should an organism commit to Stability versus Novelty? These two processes need to remain in some dynamic, competitive, and cooperative balance. The notion that opposing processes are maintained in a dynamic balance has been known for thousands of years as *dualism*, from Taoism's yin and yang. Many other examples in our own daily experience display dualism, including the experiences of Stability versus Novelty, owning versus sharing, separating versus belonging, and competing versus cooperating.

5   M. J. Russell et al., "The Drive to Life on Wet and Icy Worlds," *Astrobiology* 14, no. 4 (2014): 308–343, doi:10.1089/ast.2013.1110. See also Wolfgang Nitschke and Michael J. Russell, "Beating the Acetyl Coenzyme A-Pathway to the Origin of Life," *Philosophical Transactions of the Royal Society B: Biological Sciences* 368, no. 1622 (July 19, 2013), https://doi.org/10.1098/rstb.2012.0258.

NEOSIS

Claude Bernard, in 1865, was the first to study the way life maintains Stability in a process later termed *homeostasis*.[6] But no one has yet coined a name for the process of searching for Novelty. Michael called it *Neosis*.[7] As we show examples of how the phenomena of homeostasis and Neosis act, you can see the beginning of a grand continuity between our biological being and our emotional and conscious being, which will get us back to our original question of why we are here.

As our own thoughts evolved, our explorations around Michael's concept of Neosis led us to wonder how life internalizes Novelty. In Michael's conception, *Quintessence* is a new explanation for how living organisms capture Novelty. Michael was dissatisfied with many of the limited definitions for life because they lacked the magical qualities that we know life contains. Michael stated that, "This unified theory of Quintessence is the fifth requirement that all life forms must have to be alive, beyond metabolism, enclosing membrane, reproduction, and growth."

As species evolve, two trends emerge beyond just the mandate to reproduce. One is a trend toward increasing resource and energy flow in all forms and abstractions, including food, money, power, and information. The other trend is the openness in all species to connect with the other life around them. Since the experiences of novel chance encounters are unknowable before they happen, sometimes the fateful results of their interactions can be truly magical, be they wondrous, dramatic, loving, dangerous, or just fun.

---

6 Kevin Rodolfo, "What is Homeostasis?" *Scientific American*, January 3, 2000, https://www.scientificamerican.com/article/what-is-homeostasis/. See also the definition of homeostasis in *Wikipedia*, https://en.wikipedia.org/wiki/Homeostasis. Homeostasis is the tendency of organisms to autoregulate and maintain their internal environment in a stable state.

7 Michael decided that *Neosis* would be an appropriate term to conceptualize the search for Novelty. Neosis has been previously used to describe the uncontrolled growth of cancer cells. See also R. Rajaraman et al., "Neosis: A Paradigm of Self-Renewal in Cancer," *Cell Biology International* 29, no. 12, (December 29, 2005):1084–1097, https://www.ncbi.nlm.nih.gov/pubmed/16316756; and M. Sundaram et al., "Neosis: A Novel Type of Cell Division in Cancer," *Cancer Biology and Therapy* 3, no. 2 (2004): 207–218, doi:10.4161/cbt.3.2.663, https://www.ncbi.nlm.nih.gov/pubmed/14726689.

QUINTESSENCE

Quintessence is how species capture the Novelty of chance encounters, negotiate through opposing pressures, and grow and evolve by internalizing interactions. These processes are inherited from the basic chemistries when life first formed and continue to operate at all levels of life from bacteria to plants to animals to humans and societies. Examples of Quintessence in action include life's first formation, the creation of DNA, the evolution of simple cells to nucleated cells, the evolution of multicellular organisms, and the process whereby genetic novelty during the life cycle of a species leads to evolution. Quintessence revises the story of evolution and, along the way, humanity's own self-image.

Quintessence extends our identity from a species that just lives to survive, compete, and reproduce to a species that also lives for stimulation, connection, and creation. Our acceptance of this new identity changes not only our understanding of living with each other but also our understanding of our relations with all life forms on our planet. In the end, Quintessence helps ground us on our origins and opens our imagination to contemplate where future evolution and human culture may take us.

We have come a long way from where we started. We went from life stories to what we inherit from the earliest life forms, and then to the origin of life on Earth. Where are we going on this path? We are beginning to show the outline of the grand continuity of life forces and how they impact the meaning of our lives.

Evolution is the 2,712th most common word in non-fiction, the 9,242nd most common word in fiction, and the 3,117th most common word across all books. Evolution is more commonly used in non-fiction books.

Word Origin and History for evolution: *n.* 1620s, "an opening of what was rolled up," from Latin *evolutionem* (nominative *evolutio*) "unrolling (of a book)," noun of action from *evolvere.*

Used in various senses in medicine, mathematics, and general use, including "growth to maturity and development of an individual living thing" (1660s). Modern use in biology, of species, first attested 1832 by Scottish geologist Charles Lyell. Charles Darwin used the word only once, in the closing paragraph of "The Origin of Species" (1859), and preferred *descent with modification*, in part because *evolution* already had been used in the 18c. homunculus theory of embryological development (first proposed under this name by Bonnet, 1762), in part because it carried a sense of "progress" not found in Darwin's idea. But Victorian belief in progress prevailed (along with brevity), and Herbert Spencer and other biologists popularized *evolution*.[3]

It is curious that, although the modern theory of evolution has its source in Charles Darwin's great book *On the Origin of Species* (1859), the word *evolution* does not appear in the original text at all. In fact, Darwin seems deliberately to have avoided using the word *evolution*, preferring to refer to the process of biological change as 'transmutation'.

Evolution before Darwin

The word *evolution* first arrived in English (and in several other European languages) from an influential treatise on military tactics and drill, written in Greek by the second-century writer Aelian (Aelianus Tacticus). In translations of his work, the Latin word *evolutio* and its offspring, the French word *évolution*, were used to refer to a military manoeuvre or change of formation, and hence the earliest known English example of *evolution* comes from a translation of Aelian, published in 1616. As well as being applied in this military context to the present day, it is also still used with reference to movements of various kinds, especially in dance or gymnastics, often with a sense of twisting or turning.

In classical Latin, though, *evolutio* had first denoted the unrolling of a scroll, and by the early 17th century, the English word *evolution* was often applied to 'the process of unrolling, opening out, or revealing'. It is this aspect of its application which may have been behind Darwin's reluctance to use the term. Despite its association with 'development', which might have seemed apt enough, he would not have wanted to associate his theory with the notion that the history of life was the simple chronological unrolling of a predetermined creative plan. Nor would he have wanted to promote the similar concept of embryonic development, which saw the growth of an organism as a kind of unfolding or opening out of structures already present in miniature in the earliest embryo (the 'preformation' theory of the 18th century). The use of the word *evolution* in such a way, radically opposed to Darwin's theory, appears in the writings of his grandfather:

*The world...might have been gradually produced from very small begin-
nings...rather than by a sudden evolution of the whole by the Almighty fiat.*

Erasmus Darwin, *Zoonomia* (1801)
Use of 'evolution' elsewhere
Charles Darwin's caution, however, was futile: the word was ahead of him. By
the end of the 18[th]century, *evolution* had become established as a general term
for a process of development, especially when this involved a gradual change
('evolutionary' rather than 'revolutionary') from a simpler to a more complex
state. The notion of the transformation of species had become respectable
in academic circles during the early 19th century, and the word *evolution* was
readily to hand when the geologist Charles Lyell was writing in the 1830s:

*The testacea of the ocean existed first, until some of them by gradual evolu-
tion, were improved into those inhabiting the land.* -Charles Lyell *Principles of
Geology* (second edition, 1832)

By the 1850s, astronomers were also using the word to denote the process
of change in the physical universe, and it would inevitably become central to
the reception of Darwin's work.

'Evolution' in Darwin's theory
Once Darwin's theory had been published, to widespread debate and acclaim,
discussion was often made more difficult by the persistent assumption that
evolution must necessarily involve some kind of progress, or development from
the simple to the complex. This notion was present in the account of *évolution*
in human society by the French philosopher Auguste Comte, and it was central
to the metaphysical theories of the English speculative philosopher Herbert
Spencer. Already in 1858, a year before *On the Origin of Species* appeared in
print, Spencer was enthusiastically endorsing 'the Theory of Evolution'— by
which he meant the transformational theory of Lamarck, which Darwin's
work was set to supersede — and his keen advocacy of Darwin's theory led
to some confusion between Darwin's ideas and his own. Even now, biologists
have frequently to explain that the theory of evolution concerns a process of
change, regardless of whether the change can be regarded in the long run as
'progress' or not.

Nevertheless, despite his reluctance to call evolution by that name, Darwin did famously dare to use the corresponding verb for the very last word in his book: "From *so simple a beginning endless forms most beautiful and most wonderful have been, and are being, evolved.*"

REFERENCES

1. https://web.archive.org/web/20151117021424/http://www.forgottenbooks.com/worddata/evolution.
2. http://www.dictionary.com/browse/evolution.
3. https://blog.oxforddictionaries.com/2015/05/08/evolution-etymology/.

CHAPTER 3

Essential Tensions

"He who learns must suffer. And even in our sleep, pain that cannot forget falls drop by drop upon the heart, and in our own despair, against our will, comes wisdom to us by the awful grace of God."

—Aeschylus

The secrets of life are hidden within the incompletely explored chemistries that contain the Essential Tensions necessary for all life. We live lives of stresses and tensions. We're defining *tension* as any process that has at least one other competing state. This could be as simple as an on-or-off process that controls a simple start or stop signal, or it could entail more complicated processes that determine whether we will cooperate or compete with someone. On the chemical level, these Essential Tensions are two relatively simple on-or-off chemical states that have played essential roles in life's evolution and that we'll explore in more detail later in this book.

Many books have been written on maintaining life's balance, on balancing work and family relationships, and so on. Yet most of us are unaware of the very crucial role played by the underlying chemical balance of our sensory receptors in perceiving and understanding the world we occupy. What we don't realize is that many of these tensions are built into life's essential core. We're woefully unaware of these phenomena until they force themselves on us when we become too far out of balance. So much in life depends on a proper balance. We talk about maintaining life's balance among our tensions, but there is a great deal more to this concept than most people realize.

When we recognize these tensions, we may try to avoid them, but they often determine how we live. Many of our most cherished historical mythologies have provided us with deep insights into some of our hidden psychological and social tensions. Some of our grandest ideas concerning the tensions between those higher realms and man have descended to us as ancient mythologies of the past. These wonderful stories have provided us with valuable insights into the grand tensions between humans and fate. At life's beginning, there were also tensions that arose from the very chemical nature of the early Earth.

Essential Tensions are what drive us. The most important Essential Tensions are those that gave life both sources for energy generation and sensory molecules that sense and respond to environmental changes. These became the very first instances of a sensory system with a primitive metabolism that could respond to its environment. This was not yet life, but it was a very important beginning.[1]

What made it much more complex was the separation of these systems by membranes and the addition of other two-state chemical systems to create a self-sustaining metabolism. Perhaps the most amazing phenomenon was the ability of these chemical systems to react and adapt to their environments. Such systems, and all adaptations in life, are biochemical adaptations writ large on the tree of evolution.

There are also Essential Tensions that naturally occur when a Self is formed within a specific environment, which includes diverse interactions with other Selves and environments. It is vital for individual cells to be able to interact with their environment and other Selves just as people do. This is true whether a cell is growing by itself in a pond or is one of many cells that form a larger organism. The ability of cells to communicate through chemical signals originated in single cells and was necessary and essential for the evolution of all multicellular organisms. While the necessity for cellular communication in larger organisms

---

1   Essential Tensions may be the most primary reason that viruses shouldn't be considered alive. Viruses don't have the Essential Tensions coupled with a suitable membrane that can create chemiosmotic gradients to generate energy for a metabolism.

seems obvious, even single-celled organisms communicate with each other. Yeast cells signal each other to aid mating. Bacteria coordinate their actions to form larger complex communities called biofilms or to organize the production of toxins to remove competing organisms. These communication systems are vital to all forms of life today.[2]

CHEMICAL TENSIONS

In the larger picture, we all live immersed in a vast sea of chemistries. We are made up of chemistries working together to provide for our metabolisms, perceptions, and cognition. Almost everything about us has to do with how our internal chemistries interact with our internal and external environments. The simplest chemical tension is a perturbed chemical equilibrium with at least two chemicals linked together by a process that can change one form into the other. Some of these processes are simple, such as the attachment or removal of a proton (H+). Others may be linked with much more complicated chemical equilibria.

In most beginning-of-life scenarios, there are inherent difficulties in understanding the underlying chemical principles that led to the formation of life. We believe that an improved understanding of how chemical tensions form and function is necessary if we want to create a better scientific foundation for understanding the chemical origin of life.

Most likely, more than one chemical tension was formed with the beginning of semipermeable membranes. These membranes prevented charged chemical ions such as sodium, potassium, chloride, and phosphate from attaining their natural equilibrium concentrations, as compared to their existence in a solution lacking membranes. In addition, biophysical properties such as electrochemical gradients became structurally embedded within these semipermeable membranes. These structural nonequilibria (or disequilibria) provided energy sources that could link with the internal molecules of the first protocells (forms of enclosed, nonliving chemistry that paved the way for

---

2 *Wikipedia*, s.v. "G protein-coupled receptor," https://en.wikipedia.org/wiki/G_protein%E2%80%
   93coupled_receptor.

living organisms). The food for these early protocells would only involve their use of energy gradients formed from chemical tensions, and the accumulation of useful molecules from their environment.

In this respect, the chemistries of life become not so much open chemical systems but membrane-embedded chemical systems that are not in their normal equilibrium state. It should be stressed that becoming membrane-embedded chemical systems provided an increased Stability to these systems so that they could weather severe environmental challenges such as drying out or low temperatures. Also, they must have served as sources of energy for the first protocells. Therefore, it was important that the correct type of semipermeable membrane could form and maintain itself during protocell evolution. To maintain itself, it may have required a stable semipermeable membrane template that could form new membranes in an inheritable way. It may also have required an enzyme-catalyzed mechanism to create the molecules that became part of this semipermeable membrane.

It is also important to note that by structurally embedding these equilibrium states into cellular membranes, the first living organisms must have had the innate ability to stop and start their metabolism, much as plant seeds and bacterial spores do today. This allowed for the first instances of life to enter a dormant state and weather unfavorable environmental conditions. Just add water, or whatever was temporarily missing, to recharge their membrane-embedded Energy Flow gradients and these earliest life forms could restart their metabolism. This example of life's ability to become temporarily dormant is like a battery, drained of its electrolyte solution, that can be re-energized by adding back its electrolyte solution to create the electrochemical gradient of the battery.

Once these Essential Tensions formed, we had the potential for the ignition of life's flame. It is difficult to understand what these specific chemical systems were that subsequently led to the earliest life, but we do know that this is a basic process found in many simple chemical systems.[3] These earliest

---

3 Equilibrium shift: previously observed as a separate area of chemistry, it was not parameterized. See Peter T. Corbett et al., "Dynamic Combinatorial Chemistry," *Chemical Reviews* 106, no. 9 (August 10, 2006): 3652–3711, https://core.ac.uk/download/pdf/12925202.pdf.

tensions also produced chemical oscillations that are found embedded within several biological systems today.

ENVIRONMENTAL TENSIONS

What was the spark that began the emergence of life? *What is meant by* "life is nothing but an electron looking for a place to rest?" We hypothesize that life inherited tensions from inanimate chemistry and physics and always maintained some version of these as it evolved in complexity. Some of the Essential Tensions that life inherited from inanimate chemistry and physics include pH and ionic gradients, osmotic gradients, and electrical potential gradients.[4] The coupling of these Essential Tensions to other molecules was a critical process on the road to life and they also provided potential energy gradients that living systems could harness as an energy source for their metabolic needs.

Where do the environmental tensions in the Hadean period (4.6 to 4 billion years ago) come from? Several studies suggest that alkaline submarine hydrothermal vents are a likely source for the origin of life.[5] Since these alkaline vents were discovered less than fifty years ago, hypotheses have attempted to explain how the chemistries surrounding these formations led to the earliest

---

The concept discussed by Corbett et al. is also presented in the following literature: Hideaki Hioki and W. Clark Still, "Chemical Evolution: A Model System That Selects and Amplifies a Receptor for the Tripeptide (D)Pro(L)Val(D)Val," *Journal of Organic Chemistry* 63, no. 4 (1998): 904–905, https://pubs.acs.org/doi/abs/10.1021/jo971782q; V. T. Bhat et al., "Nucleophilic Catalysis of Acylhydrazone Equilibration for Protein-Directed Dynamic Covalent Chemistry," *Nature Chemistry* 2, no. 6 (May 16, 2010): 490–497, doi:10.1038/nchem.658; Richard G. Lanzara, "Method for Determining Drug Compositions to Prevent Desensitization of Cellular Receptors," US Patent, US5597699A (September 30, 1992), https://patents.google.com/patent/US5597699; Irving R. Epstein and John A. Pojman, *An Introduction to Nonlinear Chemical Dynamics: Oscillations, Waves, Patterns, and Chaos* (Oxford: Oxford University Press, 1998).

4   For a wonderful introduction to the biochemistries of life, see Nick Lane, *Life Ascending* (New York: W. W. Norton & Company, 2009).

5   Nick Lane, "The Vital Question: Energy, Evolution, and the Origins of Complex Life," (New York: W. W. Norton & Company, 2016). See also Wolfgang Nitschke and Michael J. Russell, "Beating the Acetyl Coenzyme A-Pathway to the Origin of Life," *Philosophical Transactions of the Royal Society B: Biological Sciences* 368 (2013), https://doi.org/10.1098/rstb.2012.0258 (Michael's favorite); Nick Lane and William F. Martin, "The Origin of Membrane Bioenergetics," *Cell* 151, no. 7 (December 21, 2012):1406–1416, doi:10.1016/j.cell.2012.11.050, https://www.cell.com/cell/fulltext/S0092-8674(12)01438-9.

primitive life. What these hypotheses have in common is that the early ocean was acidic, while the deep-sea vents ejected an alkaline effluent. The surrounding walls of these vents were formed of carbonaceous precipitates with small pores that separated the warm alkaline fluid from the cooler, acidic seawater. This interface between the two environments created a natural pH and charge gradient that could have driven chemical reactions toward more complex organic molecules, setting the stage for the formation of early protocells and, ultimately, a living cell.

How Might Biological Energy Systems Have Originated and Evolved?

In 1961 Dr. Peter D. Mitchell proposed his chemiosmotic hypothesis. At the time, this was a radical proposal that was not well accepted. The prevailing view was that the energy of electron transfer was stored in a stable chemical intermediate. The problem was that no chemical intermediate was ever found. As the evidence for proton pumping through the complexes of the electron transfer chain grew too great to be ignored, the weight of evidence began to favor the chemiosmotic hypothesis. In 1978 Dr. Mitchell was awarded the Nobel Prize in Chemistry. Evidence for this method of energy generation has been found to occur in all forms of life.[6]

As Dr. Lane describes in his wonderful book, *The Vital Question*, chemiosmotic coupling gives life a huge range of metabolic versatility and potential chemical space, allowing cells to eat and breathe practically anything. Even such disparate organisms as plants and trees have mitochondria, just as we do, which transfer electrons down multitudes of these respiratory chains pumping protons across membranes. These moving electrons (electrons move in the opposite direction to protons to balance the charges) and protons create the chemical gradients that have sustained you from the womb: you pump 10^{21} protons per second, every second. Your mitochondria were passed to you

---

6   *Wikipedia*, s.v. "Chemiosmosis," https://en.wikipedia.org/wiki/Chemiosmosis. See also Peter Mitchell, "Coupling of Phosphorylation to Electron and Hydrogen Transfer by a Chemi-Osmotic Type of Mechanism," *Nature* 191, no. 4784 (July 9, 1961): 144–148, https://doi.org/10.1038/191144a0.

from your mother, in her egg cell, the gift of giving that goes back for countless generations to include that first stirring of life over four billion years ago. Death is the ceasing of electron and proton flux, the final establishment of equilibrium, and the end of these Essential Tensions. If life is nothing but an electron looking for a place to rest, death is that electron come to rest.[7]

Another key Essential Tension that might have provided the earliest life with an electromotive force and is also important in the initial biochemical events involving the fertilization of the human egg cell is the Na+/H+ (sodium/ proton) exchanger that occurs across lipid bilayers and cellular membranes. The proteins that sustain this function are universal in prokaryotic, animal, and plant biology. They provide the electromotive force that arises when this specific Essential Tension occurs at the cellular membrane.

The electromotive force describes the separation of charged ions by a cell's membrane that gives a charge separation and thereby an electromotive force as a voltage to a cell's membrane. This electrification of the membrane may have been an essential step on the chemical road to life. These Na^+/H^+ exchangers are expressed in the plasma membranes of mammalian cells where they play multiple roles in cell pH, sodium transport, volume homeostasis, cell motility and as a platform for signaling complexes. The gene coding for Na+/ H+ exchangers has been found to be similar in everything from the simplest prokaryote to the most advanced multicellular eukaryotes.[8]

Living organisms use energy sources that are generally conserved across all life. The essence of all living metabolism involves combinations of chemiosmotic and reduction/oxidation (redox) chemistries. These redox chemical reactions are found paired in almost every form of life. This is due to the electron flow through the enzymatic redox centers of a living organism's metabolic

---

7   The above paragraph is paraphrased from Dr. Nick Lane's book *The Vital Question* (2015) ISBN-13: 978-1781250365.

8   Povilas Uzdavinys et al., "Dissecting the Proton Transport Pathway in Electrogenic Na^+/H^+ Antiporters," *Proceedings of the National Academy of Sciences* (PNAS) 114, no. 7 (February 14, 2017): E1101–E1110, https://doi.org/10.1073/pnas.1614521114. See also Povilas Uzdavinys, "Establishing the Molecular Mechanism of Sodium/Proton Exchangers," PhD thesis (English), Stockholm University, Faculty of Science, Department of Biochemistry and Biophysics (2017), https://pdfs.semanticscholar.org/cb19/ da611842ce0f9fbcd24b4cf67c44ed6d03f2.pdf.

enzymes. Forms of metabolism, such as photosynthesis in plants and respiration in animals, turn out to be basically the same in the chemical sense that they involve the transfer of electrons down these respiratory chains of linked molecules. All life is driven by redox chemistry using similar respiratory chains of protein/enzyme complexes. This is very much like the flow of an electrical current down a wire. This is what happens in cellular respiration, which explains *Albert Szent-Györgyi's comment that* "life is nothing but an electron looking for a place to rest."[9]

SENSES

We have five basic senses of sight, hearing, smell, taste, and touch, but how did earliest life begin to sense and respond to its environment? Our working hypothesis is that these processes began as relatively simple chemical systems in life's early evolution. Essential Tensions initiated the first metabolism and sensors for protocells to grow, adapt, and respond to their environments, which eventually led to the first life.

An organism's ability to detect and monitor its environment arose through the Essential Tensions imposed upon its chemical sensors. This requires the concomitant development of sensory molecules on the surface of the cell membrane, which provide information (feedback) to the cell's internal metabolism. We're only at the beginning of our discovery of these intricately linked systems. We are trying to make sense of billions of years of evolutionary history! The complex web of interconnected pathways of signaling molecules and metabolic enzymes within any living organism makes our attempts to describe them seem comically crude. We should appreciate that we're at the very beginnings of these endeavors and not despair that we haven't figured it all out by now.

Without senses, an organism would be blind to its environment and any other organisms. Being able to sense one's environment may be one of the

---

9   Electrons are separated from sources of energy such as food. They are then transferred through a series
    of intermediary proteins to an oxygen atom and two hydrogen atoms to form a simple water molecule
    (H_2O).

most important descriptions for life, yet it is one of the least appreciated and understood. All life requires a way to sense its environment to detect danger or reap rewards. Life responds to its environment with inherited biophysical mechanisms that preserve homeostasis or allow it to flee stress, pain, or danger. These responses rely on the living organism's ability to recognize these potential dangers through biological sensors, which are all, basically, chemical sensors.

To respond to external stimuli, cells have developed complex mechanisms of communication that receive a message, transmit that information across the cell membrane, and produce changes within the cell in response to the message. In multicellular organisms, cells continuously send and receive chemical messages to coordinate the actions of organs, tissues, and cells. The ability to send messages quickly and efficiently enables cells to coordinate and fine-tune their biochemical functions, which are critical processes for all life.

We mentioned that these processes began as relatively simple chemical systems in a perturbed equilibrium, which is sometimes called a disequilibrium. Such chemical systems have a built-in feedback mechanism as explained by the mathematics of the net shift (Appendix III C). In biochemistry, toxicology, and pharmacology, this negative feedback is variously called habituation, autoinhibition, or tolerance. It is a necessary component for all life and is naturally built into an organism's responses to environmental stimuli. Understanding habituation at the very basic level of sensory receptors may help us understand the molecular mechanisms behind our cells sensing and responding to environments and may eventually aid our understanding of memory and learning. A potential physical link to understanding how we sense and respond to environmental stimuli is a well-known physiological law called Weber's law, also known as the Weber-Fechner law (Appendix III C). That Weber's law can originate from the description of a net shift within a poised chemical equilibrium or the physical laws of a simple balance allows us to see deeper connections between the world of our everyday senses and the physical-chemical world. This provides a fundamental link to understanding an important aspect of all life, which will become more apparent when we consider the equations of balance and how they relate to chemical equilibrium and the net shift.

RECEPTORS

The signals, which are often chemical, are sensed by membrane molecules called receptors that enable cells and whole organisms to react to changes in their surrounding environment with suitable responses. Receptors are numerous. -All life has receptors. Most receptors that scientists have studied interact with a select number of molecules that transmit signals into the much more complex molecular milieu within the cytoplasm of cells. Receptors are responsible for most biological processes including all responses of an organism's sensory system, reproduction, growth, development, behavior, emotions, and pain. They are also critical regulators of an organism's immune system and are directly involved in host defense against invading pathogens.

Recognizing the fundamental principles behind such chemical behaviors help us to understand how our own senses and receptors detect the perceptual realities of the world. These perceptions can often transmit false or confusing information, which may occur as illusions or hallucinations. A whole book could be written just on receptors.[2] Each of our five senses (smell, taste, touch, sight, and hearing) have in common these molecules called receptors that interact with the external environment.

Receptor molecules, which have evolved naturally from basic chemistry as chemical- systems, have several remarkable features that allow them to respond to their environments when perturbed. One remarkable feature is that they appear to be poised in a two-state chemical equilibrium with one state functioning as the 'on' state and the other as the 'off' state. This allows them to sense environmental changes through an equilibrium shift, or net shift and transmit these changes into the interior of the cell. It is fascinating that chemical systems can acquire receptor-like traits from this relatively simple chemistry.

The poised chemical equilibrium of a receptor is an Essential Tension that exists between 'on' and 'off' chemical states and allows us to sense the world and the universe. The behavior of this equilibrium also contains many surprises because it acts in a nonlinear way (i.e., a graph of the stimulus and response of the sensory system would not be in a straight-line relationship). This means that if the stimulus is larger, the response doesn't increase in a straight line. For all our senses and receptors, the response to larger stimuli often decreases, which is a built-in component of the underlying chemistry of our receptors and senses (if you wish to go down the deeper rabbit holes, see Appendix III A–C).

You are currently using these receptors to read these words right now. If your eyes were exposed to a very bright light, you wouldn't be able to see these words right away, but eventually, your eyes would readjust. Your sensory receptors' decreased response to a large or stronger stimulus is an example of negative feedback, which is sometimes called fade, or desensitization, or any other of a large variety of scientific terms (see Appendix III A). While you are reading this, you are also using other receptors in your brain to process and understand what you're reading. Your brain is full of nerve cells that have multiple receptors that process the multiple sensory and physiological processes required for you to live. Our sense of our own consciousness arises from these brain cells talking to each other through their receptors. These processes can be interrupted by some very simple molecules that interfere with the activation of cellular receptors, some of which are useful, such as anesthetics and medical drugs, and some of which can be harmful, such as poisons. Such molecules demonstrate that our consciousness can be environmentally manipulated quite easily by relatively simple molecules.

CHAPTER 4

Quintessence

by Dr. Michael Kuperstein

"One learns to look behind the facade, to grasp the root of things. One learns to recognize the hidden currents, the prehistory of the visible."
—Paul Klee, artist

W hen I was thirteen, I had a fateful encounter with death. It was the summer of 1967, and my mother decided that our family should vacation in Atlantic City for a week. When we arrived there for the first time, my brother and I couldn't wait to hit the beach. It was summer Mecca: the smell of the boardwalk, the glitz of the trinket shops, the sounds of the merry-go-rounds, and the beautiful people walking in lazy shuffles. I had just learned to swim two years before, and I galloped into the ocean, kicking and yelling. My mother warned me not to swim out too far. She said that she wanted me to see her at all times under her beach umbrella.

Body surfing was thrilling, and I went out a little further to catch the bigger waves, but on that day there was an undertow that came without warning. I didn't even know what an undertow was at the time. As I looked up from time to time, I noticed that the beach looked farther and farther away, and I had no idea why. I would later learn that the lifeguards were blaring warnings of the undertow, but I never heard them. In a few minutes, the wave swells became larger and larger, and then, I knew that I was out too far, so I started to swim hard toward the beach. But for all my efforts, I was going out to sea even farther. I was near panic. I saw four other people, about thirty yards away, also struggling. I got my myself together and started to float on my back to conserve energy. I felt oddly calm in the thought that people on the beach must surely know that the four of us were stranded out there, and all we had to do was wait it out until help arrived.

After a long half hour that seemed like half a day, a motorized rubber raft did in fact come to pick us up and bring us in. My mother was furious with me for not doing as I was told and ordered me back to our hotel. I think she was more angry that I had put her in a high degree of anxiety and worry during the whole episode than anything else.

On the way back to hotel, I walked past a man in his fifties, lying still on his back on the beach, about ten feet from the water. He had a large bear belly and fat neck and he wore dark blue swim trunks that were too tight. He didn't move while a man and woman held him below his neck and turned him over. Water gushed out of his mouth. After a few minutes his skin started turning a shade of blue. I wanted to turn away, but I was glued to watching this scene from about twenty feet away. The man and woman started screaming, and more people came by. Soon, an ambulance rushed its way onto the beach, sirens sounding, and medics ran to the man. I could hear a lot of commotion on their walkie-talkies. In minutes, the man died. I could see the distressed helpers giving up. My fingers and toes grew extremely cold and my stomach started heaving. It was all I could do not to vomit right there on the beach. My mind went numb. For the first time, I witnessed a person die in front of me.

I woke up multiple times at night for the next three days, visualizing the beach scene over and over again, and as the weeks went by, I tried to analyze why I felt the way I did at witnessing death. I kept asking myself the same questions again and again: What does it all mean? What happens after you die? What is the value of living? I asked the same questions that people all over the world, throughout history, have asked: where do we really come from and why are we here?

The first known record of such questions was made by the Babylonians 3,800 years ago. The Babylonians wrote their creation myth on seven tablets named the *Enûma Eliš*. As all people do, they built up a story that made sense to them from what they experienced from day to day. You can get an indication of what they knew and believed from the first lines on the first of those tablets.

"When on High"

When the sky above was not named,
And the earth beneath did not yet bear a name,
And the primeval Apsû, who begat them,
And chaos, Tiamat, the mother of them both,
Their waters were mingled together,
And no field was formed, no marsh was to be seen;
When of the gods none had been called into being.

This text goes on to describe the sea, earth, sky, and wind as the four fundamental elements of our world. And so, here begins our human need to bring order to what we experience through categorization and organization. We need to deconstruct our world into fundamentals and processes so as to reconstruct meaning. It gives us some comfort to feel that there is some foundation, some basis, that might explain the world we live in and journey we take in it.

QUINTESSENCE

People believed in the Babylonian fundamental elements for the next 1,300 years until the time of the Greeks, about 2,500 years ago. The Greeks then came out with their own version of the four fundamental elements: air, fire, earth, and water. Then, around 2,350 years ago, Aristotle added a fifth element, which he called Quintessence, also called ether. For Aristotle, Quintessence is the material that fills the region of the visible sky above Earth.[1]

To get a sense of how universal these concepts were at the time, you can compare them with those of other cultures. The ancient Hindus, Buddhists,

---

1   *Oxford Living Dictionaries*, s.v. "Quintessence," https://en.oxforddictionaries.com/definition/quintessence
Noun
    1.   The most perfect or typical example of a quality or class.
       "He was the quintessence of political professionalism."
    1.1. The aspect of something regarded as the intrinsic and central constituent of its character.
       "We were all brought up to believe that advertising is the quintessence of marketing."
    2.   A refined essence or extract of a substance.
    3.   (In classical and medieval philosophy) a fifth substance in addition to the four elements, thought to compose the heavenly bodies and to be latent in all things.

Japanese (Godai) and Tibetans (Bön) developed similar elements. And the Chinese (Wuxing) had their wood, fire, metal, and water. These categories stayed about the same in our cultures until the Late Middle Ages/early Renaissance eras.

In Aristotle's system, Quintessence had no qualities: it was neither hot nor cold, neither wet nor dry, and it was incapable of change, with the exception of change of place. By its nature, it moved in circles. Medieval philosophers granted Quintessence changes of density, in which the bodies of the planets were considered to be denser than the medium that filled the rest of the universe.

Even in the early 1900s, scientists still considered the concept of the aether as perhaps the thing that constitutes space itself. Then physicists Michelson and Morley disproved it with their experiments using light traveling in different directions to measure possible effects of the aether, of which they found none. Even to this day, theories are put forward on what could be the structure of a space vacuum. It seems difficult for us to think of nothingness. Just giving it a name gives it somethingness.

So, after the ages, Quintessence has taken on the meaning of some kind of medium that connects everything else and in which all else moves and interacts.

Before the rise of the science of chemistry, none of the elements included living organisms, so only recently did people even consider any connection between what makes up life and what makes up nonlife. The idea that life is made up of a subset of chemistry is only a few hundred years old.

Looking back at all this today, we might wonder what they were thinking! Based on what we know today, they were so way off, but today we are still asking the same questions—just in a more nuanced way: Where did biological life come from and why is life here? Will our answers today hold up through the next 2,000 years of human learning?

The journey I am starting with you here is about another Quintessence. This one is not about a medium for the elements of matter but about what connects the processes of life.

Today, the common answer to the question of what constitutes life is that life is made up of four qualities: some metabolism, in various cell enclosures, that grows and reproduces/evolves. But does this really get at how I or you feel about being alive? Not even close! These four life qualities feel like a recipe. I have watched several documentaries maintaining that, someday, factories will churn out man-made life to order, with these four qualities, much as car factories churn out cars.

So what do metabolism, cell walls, growing, and reproducing have to do with the magic of being alive or how we feel being alive: tasting your mom's best dish, a first kiss, looking into your baby's eyes, the memory of a smell from childhood, seeing the Grand Canyon for the first time, reaching a mountain summit, crossing the finish line of a marathon? Not much. We laugh at the Greek concept of the classical elements, and in 2,000 years, people will probably laugh at the four qualities of life we talk about today. There is no question that these four qualities are necessary, but they don't really satisfy as the fundamental explanation of what we experience as the magic of life and the processes that all life, in our world ecosystem, share.

And so, this story is about the search for a fifth quality of life I call Quintessence, which, I claim, is the set of Essential Tensions between opposing pressures that all life in our ecosystem must have, and at all levels of life from bacteria to plants to animals to us to our societies. Once I explore this type of Quintessence with you, I would like to show how it can shift our entire understanding of how life grows and evolves. And in the end, it can begin to suggest new meaning to the questions of where we came from and why we are here.

Some years after the beach experience, I came face to face with Darwin's ideas on evolution for the first time. I was seventeen and the year was 1971. This was a time before the Internet, before cell phones, before the personal computer, before fax and even before the microwave oven became popular. The year 1971 was when Walt Disney World opened, the voting age was lowered from twenty-one to eighteen, John Lennon came out with the song "Imagine," the Supreme Court upheld busing for desegregation, and Intel introduced the microprocessor.

That year, I took my first computer course at the Norwich Free Academy in Norwich, Connecticut. Our "computer" was a fancy calculator called the Olivetti and was the size of a fat briefcase. You could type in simple statements on its keyboard, and it would run them in sequence with decision points and loops, producing a small paper print-out from a roll just as old cash registers did. I was fascinated by the world this opened up for me. It gave me a feeling of power to control a sequence of commands that did things such as compute the mathematics number *pi* to a hundred decimal places and print out funny pictures on the little printer.

My teacher, who also learned this for the first time that year, told me about the school's computer, which was located next to the principal's office. It was in the only white building on the edge of our large campus. I pushed him to show it to me, and he did. I learned that the vice principal controlled it. So, a few days later, after school, I met with the vice principal and asked him if he could teach me to operate the computer after school.

He said, "Why don't you read the manual and learn it yourself?" He told me the computer was turned off after school, so I would have to turn it on and feed it programmed punch cards. When I was done, I had to turn it off. He agreed to give me the key to the room if I would keep him informed about what I was doing. Well, after that, I would stay there for hours on end for the next year and half until I graduated from high school.

The school's computer was a used IBM 1401, the size of a big refrigerator in a room ten feet wide by ten feet long. Next to the computer was a large line printer that could print five lines per second. This was astounding to see at the time. Hugging the printer was a huge key-punch machine and an automated electric typewriter. It was, actually, an old-fashioned manual typewriter hooked up to the computer by some sort of custom control. I was in heaven. With all the equipment, there was just enough room for three to four people to stand. But when I was there, I was the master commander of all things computing.

I spent a year and a half in that room, averaging four hours per school day. I was fascinated by how this powerful machine could be started with a single punch card that was called the boot card and had about fifty micro-instructions on it. The manual had a boot-strapped set of instructions (the origin of the term *booting up* when applied to a PC). The boot instructions created new instructions, which then created other new instructions, and so on, until a complex loop was set up to prepare reading instructions from the rest of the punch cards that composed my program. From the initial fifty microinstructions, about 1,000 instructions were created, mostly through the computer's expansion and modification of its own instructions.

This got me thinking about the essence of where this power came from. What was the nature of this essence? Could this essence be used to understand how brains work? Could the essence of how the computer modified itself be used to understand how learning happens or how life evolves? In the same year that I was consumed with the school's computer, I took my first biology course in which Darwin's ideas on evolution were taught.

Those two experiences prompted me to ask a question that seemed simple enough, but whose answer I could not find no matter whom I asked or where I looked. It was a loose thread that troubled me about Darwin's ideas. I kept pulling and pulling on that thread for three decades as some of the most accepted ideas on evolution began to unravel in my mind. And through a re-knitting process, I would come to a new understanding of what might really drive Growth and evolution in biological life.

DARWIN'S CONCEPT OF EVOLUTION

Let's start with what I was taught about evolution and what has become dogma in the latter half of the twentieth century in the biological community. In its most simplistic form, Darwin's theory of natural and sexual selection goes like this: All animals produce offspring that have traits similar to those of their parents but that also have random variations of traits that differ from their parents. If the offspring survive and mate successfully, as adults, they will reproduce another generation that carries on most of their traits and, again, some other random variation of traits. As the environment changes by climate, predators, and food supply, the source of novel ways to adapt and better survive these changes lies in the random variations that each generation inherits. Then, by virtue of the animals' success in surviving and mating, successive generations of animals will fit better with their environment than did the previous generations. Those traits that enable animals to survive and mate well will be enhanced over generations, while those traits that hinder survival or mating will be diminished.

The traits of those animals surviving successfully over many generations spread to a larger and larger percent of the species population. In this way, animals are said to evolve. According to this process, all sources of change for evolution to occur must come from the random variations of traits that each generation inherits. Those offspring that best fit their environment and mate will continue to evolve, and those that die or don't mate will not. The famous phrase "survival of the fittest" has its origin in natural and sexual selection. When early apes were forced by a changing climate to leave the African jungle and enter the savannah, the selective pressure of the new environment's challenges to survival caused the apes to evolve into humans.

The process of natural and sexual selection would more accurately be described as the "extinction of the unfit." A selection process is subtractive and passive rather than active. However, the positive side of selection is that it is, indeed, natural since there is no agent behind the scenes, deciding who is selected and who is not.

The often-used term *selective pressure* is just a bad choice of words and meaning. In physics, pressure comes from a force. There is no force in selection. When the term *selective pressure* is used in relation to evolution, the real meaning of the term is "hardship to surviving." A changing environment such as the one the apes experienced in the savannah made finding find food and surviving more difficult. The whole set of conditions that allowed the apes to thrive in the jungle was changed. In the savannah, the apes had to hunt differently, move differently, avoid predators, and withstand the elements differently.

So, the only thing that increasing the selective pressure does is increase the threshold needed to survive and reproduce. Because selection is a passive process, less of the population survives. How does the passive nature of selection and the passive nature of random variation improve or speed up evolution, which is where the term *selective pressure* is applied? Well, they do so indirectly. According to this theory, as the survival hardships increase and the survival rate decreases, the variation of traits among survivors may increase the proportion of the next generation that has the new traits, allowing these offspring to cope with the increased hardships. The theory holds that because there are fewer offspring lacking the adaptive traits, there is less risk of watering down those new, more adaptable variation traits, thus increasing the rate of evolutionary improvement.

But this assumes that the rate of the variation of traits is fast enough to cope with the rate of increasing hardships in a changing environment. Otherwise, the whole species becomes extinct.

An easy way to understand this issue is by analogy. Suppose I represent the combination of traits coming from two parents, each parent having some combination of a hundred balls of four different colors (red, yellow, blue, and green) representing the different traits that can be inherited. The offspring will inherit some random combination of these colored balls, which can be represented as 200 balls composed of 50 red balls, 50 yellow balls, 50 blue balls, and 50 green balls. Then, suppose I place all 200 balls into a bucket out of which I choose 100 balls as the traits (colors) representative of one offspring. Clearly, as I repeat this process, I am going to get a different combination of traits each time, with little chance of getting the exact same number of balls of each color. Each sample of balls represents a random variation of traits of an offspring relative to its parents. Each offspring uses its traits to help it fit in its environment and mate. And through a Darwinian process, successive offspring can evolve to better fit with their environment. The catch is that this process works only up to a point. So far, we have required only 100 items to be taken in each sample and only ball-shaped items of only four colors. What if the environment changes so much that no combination of 100 colored balls with the four colors or traits will allow the offspring to survive? Then, the species becomes extinct. More importantly, in this analogy, the species is defined by the traits of the hundred balls with the four colors, so there is no way to understand how it could evolve into a new species as represented by other solids such as cubes or cylinders. In this analogy, there is no way to evolve to have more balls or more colors—different traits—any combination of which could represent a new species.

CONSEQUENCES OF DARWIN'S IDEAS

On the one hand, Darwin's theory was simple, general, and elegant, which made it very appealing. On the other hand, it couldn't explain how new species evolved. This troubled both Darwin and other biologists, and over time, as the structure and meaning of DNA was discovered, biologists formed a revised theory called the modern synthesis. Biologists had discovered that DNA is not always reproduced perfectly and changes in the DNA (called mutations), during the lifetime of organisms, are transferred to the next generation. DNA mutations, then, became the answer to evolving new species, since by analogy with the hundred colored balls, the mutations enabled new, random variations in number, color, and shape of the traits that the offspring could have.

So, then, there were two types of random variation of traits that were used. One was a random sampling of offspring traits from their parents and one was a small random change of the DNA that could propel one species to evolve into another through evolution. The modern synthesis theory holds that variations are blind and only genetic (based on DNA), and that events during the life of the organism do not contribute to evolutionary change.

Today we know a lot more about our genome and how it changes than I did in high school. We know, from peer-reviewed journal publications on more than a hundred species that variations are neither blind nor only genetic, and events during the life of the organism may contribute to evolutionary change.

There are about three billion combinations of the four fundamental DNA building blocks, or bases—let's call them A, C, T, G, for short—in the sequence of DNA in the human genome. About 2 to 3 percent of our roughly 25,000 genes produce our protein, and the rest manage how, when, and where the proteins are used in the tissues of our body as we develop and survive. With respect to the variations assumed, much of the variations are random combinations of traits from our parents, while just a tiny amount are random changes, or mutations, of individual bases.

I was more troubled by another consequence of Darwin's ideas on evolution. Because of my exposure to how computers can improve their function by modifying their own code, I thought that, maybe, some type of active self-modification might be a good analog to both how the brain learns and how life evolves. I could not reconcile the computer's active modifications of computer code on itself—which is, in part, where the computer's power comes from—with the passive nature of Darwin's theory of natural selection and sexual selection.

Who cares if evolution is competitive, random, and passive or if evolution is also active and includes cooperation? To me, there is fundamental issue here of our identity as an organism, the place that we and all of life around us have in our ecosystem, and where we all fit together. Are we and the animals around us just nature's automatons driven by selfish genes and required to compete to survive and mate as the bargain for why we are alive? Or does our being alive matter to what happens to the next thousand generations of evolution and the life around us?

I'm not talking about world leaders or leading countries that can change the world's climates or food supply and thereby change the course of both human and animal evolution. I'm talking about us as members of the *Homo sapiens* species interacting with the evolution of all of Earth's species. After all, all life shares the qualities of metabolism, cells, reproduction, and evolution. Are there other processes that all of life share?

In Darwin's theory, the random production of varying traits is at the heart of novel traits that are passed through the selection sieve. If novel traits are truly random, then they are totally independent of development and totally independent of any behavioral interactions that occurred during their lifetime. This leads to the conclusion that the only purpose for living is to survive and reproduce, given the cards we are dealt from our parent's traits and some other random cards. From this perspective, how we live has no effect on what our children get from us and how we evolve, as long as we just survive and mate.

On the other hand, if anything that happens to us while we live has any positive gains for our children in terms of the traits they get from us, then how our species chooses to engage with our environment can make a difference in how we evolve. From this perspective, our purpose in living must include what we do besides just surviving and reproducing.

The crux of the difference between these two perspectives is all about where novel traits come from. From what I learned about computers and what I was taught about Darwin's theory of evolution, my instinct was to go for more of an adaptive approach to evolution, and so, my journey started. I needed to examine the reality of a purely random variation from what we know about genomic data and if there were a problem with it, I would need to put forth an alternative theory not based on pure random variation.

For each of us, there are moments of contact and discovery that change everything: a chance meeting that begins a long friendship, a first hello that leads to a marriage, a first stage performance that catches fire with its audience, or an influential book that changes a culture. And then we understand that life doesn't happen by plans alone. Connection and reaction are key sources of something new that can grow and evolve. Why should this only be true of the human experience? Why can't it be true for the polar bear in the arctic, the bird of paradise in the jungle, the sunflowers in our garden, the algae in the ocean, and the bacteria in our gut? Why can't all life be ready to connect and react to a Fateful Encounter that leads to Growth and evolution?

Other than what we do to survive—the habits of eating, work, sex, shelter, and security—what makes one encounter take hold while so many others fade? In our lives, we may be ready for something, primed to accept, prepared to react. It may not be something we thought about but something that just happens with a gut reaction. It could be something we were unaware that we were missing. But when the right chance encounter happens, there is some chemistry, a fit that brings a natural increase of stimulation or energy to our lives. It is palpable. Our attention sharpens, our heart may beat faster, and the experience may feel other-worldly. It may take on the illusion of truth. For a moment, it may even transcend doubt, combining feeling and thought into certainty. And by it, we are changed. We are turned. We become aware. We grow.[2]

---

2   For more of Michael's notes on Quintessence, see Appendix IV.

CHAPTER 5

Stability versus Novelty

"All my life through, the new sights of Nature made me rejoice like a child."

—Marie Curie

The following final chapters are only sketches of our discussions. Hopefully, I have faithfully preserved the essence of our lines of thought.

Our lives are composed of both Stability and Novelty. We lead lives of relative Stability enmeshed in a dynamic order. We have our routines and regular cycles of activities. Occasionally, something new happens that introduces us to Novelty. We might learn something new or understand something that we failed to see before. Our basic routines may not change, but we become more aware and enriched by these experiences. This Novelty provides us with further opportunities to expand and change our lives. It is what we define as the incremental changes affecting the underlying, stable chemistries of life.

Life is opportunistic. After eons of evolution, life has discovered that to have Stability it must also search for Novelty. From life's perspective, this makes sense because life has evolved in a world that has both dangers and rewards. Sensing dangers, life has the option to seek Novelty for potential rewards while avoiding potential dangers. Michael suggested that the process of searching for Novelty be called *Neosis*. He saw the beginning of a grand continuity between our biological being and our emotional and conscious being. This suggests a fundamental view of our basic psychology as the product of billions of years of evolution. Life itself creates an increase in potential Novelty. New connections emerge between life's chemical elements and the surrounding matter and energy. At the beginning of life on Earth no one could have predicted that we would discover how to transform matter into energy via nuclear fission or fusion. As scientists, we are curious. We seek out Novelty to study and understand it because it is inherent within us.

When the first chemical reactions toward life began, there were also Essential Tensions between Stability and Novelty. There developed a stable, core set of reactions that could also react with other chemicals within this initial incubating environment. This became the first metabolism. It preceded the first protocell formation and is therefore not likely defined as life, according to our five essential concepts. However, it also initiated the development of the other essential concepts necessary for life to develop.

HOMEOSTASIS

The balance between Stability and Novelty is what creates the overall balance in our lives, called homeostasis by the medical physiologists referring to our metabolism. Homeostasis describes those linked chemical networks that balance an organism's metabolism and Growth. Walter Cannon used the term *homeostasis* in 1930 while referring to how the body maintains its temperature, and other key physiological variables. The term was first mentioned in his book *The Wisdom of the Body*, which describes how the human body maintains steady levels of temperature and other vital conditions, such as the water, salt, sugar, protein, fat, calcium, and oxygen content of the blood. Homeostasis derives from the Greek words for "same" and "steady," and refers to any process that living things use to actively maintain the physiological conditions necessary for survival. Homeostasis in a living organism is the core *balance* among energy sources to maintain Growth and proper Energy Flow. In general, this involves the biochemical balance between an organism's use and disposal of energy and the biochemical networks that the living organism uses to grow and, eventually, reproduce. This must also include the disposal or recycling of waste in the forms of energy, chemical species, and excess metabolic capacity.

The relationship between homeostasis and cybernetics is derived from how linked networks are controlled. The science of cybernetics (from the Greek for "steersman") was defined in 1948 by the mathematician Norbert Wiener as "the entire field of control and communication theory, whether in the machine or in the animal." Cybernetic systems can remember disturbances and thus are used in computer science to store and transmit information. Negative feedback is a central homeostatic and cybernetic concept, referring to how an organism or system automatically opposes any change imposed upon it. In chemical and biological systems, this is reflected in the fact that these systems can be perturbed by environmental factors that shift the underlying chemical equilibrium into a perturbed state that then resists further changes. This creates the reduced responses associated with larger stimulus states such as habituation or desensitization in chemical and biological systems.

Now, for the first time in history, humans have gained the power to manipulate and control their biochemical adaptations using science, which will usher in new eras of emergent tensions. Life may look very different than it does now due to its innate abilities to adapt and evolve. The Essential Tensions, with their potential for net shifts acting upon their poised chemical disequilibria, led to the most basic development of life and evolution by allowing life to sense and adapt to environmental changes. The amazing beauty of life is that the basic Essential Tensions have coupled with many other chemical reactions of life for billions of years. Throughout this long history, life has discovered ways to utilize and carry these Essential Tensions forward to today.

Change must precede evolution but is also an essential part of it. Change is the Novelty sought by organisms searching their environments. We should examine some of the ways change occurs and how it might become embedded within life:

1. *Sources and drivers of change.* Fateful Encounters and Novelty drive changes through attractive interactions that may facilitate survival.
2. *Regulation of change.* Stress and pain drive change through feedback in order to maintain survival.

3. *The potential for change.* Energy is the currency of survival and change.
4. *Context of change.* These contexts are change within Self, change within a group, and change across peers (i.e., cooperation versus competition).
5. *Propagation of change.* This involves turning copying and communication into inheritable material and feedback loops in metabolism and sensory receptors.

There exist fundamental tensions between Stability and Novelty, which can lead to change. Change driven by novel external interactions needs to happen to enable overall Growth and adaptation. But if the external drivers of change are too great, then the organism may die from instability since change without regulatory control is unstable. So, there exist tensions between Novelty and Stability and neither can totally dominate, but both are in a continuous tension that eventually achieves the necessary balance between the two.

These fundamental tensions between Stability and Novelty find a dynamic balance in all living organisms, and this dynamic balance is, fundamentally, much more complex than the two-state balance and its relationship to senses and energy gradients across cell membranes (discussed in chapter 3 and the appendix on Essential Tensions). We can scarcely begin to imagine the complexities of these metabolic systems within living cells. They're the culmination of billions of years of evolution! We don't often consider the weight of such chemical complexity over such a long span of time. We have barely begun to probe these systems that are within all of us. They represent the Novelties and Fateful Encounters that have occurred during life's long journey, which have been successfully incorporated into our current metabolisms.

The chemical evolution of life consists of those processes that include the biochemical and genetic mechanisms of all life. Life has been building onto those basic chemistries for more than four billion years. From the moment life first emerged, the potential for interactive chemistries has also given birth to Novelty. This subset of basic chemistries has undergone many changes and additions. The caveat is that they've changed yet remain embedded within our own biochemistry today. This is analogous to our description of the Wright brothers' first airplane flight. How much of the Wright brothers' original engineering elements are found in our planes today? Probably not much. These changes have been the products of the dynamic balance between Novelty and Stability coupled with Fateful Encounters to create the planes of today.

It was life's earliest chemistry that captured Novelty and transformed it into Stability. This was at least in part accomplished by linking together networks of chemical equilibria into semi-stable systems that could expand and grow. This linking together stabilized these networks. These incredible molecular systems are what keep us alive, and they're poised to undergo more evolutionary change. As beings emanating from these molecular underpinnings, we're on the cusp of newer Novelties that could radically transform our future existence. These Novelties aren't necessarily the latest scientific breakthroughs, as reported in the news. They are developments that aren't on our radar. They might arise from the fact that more and more of us consume supplements and drugs. They might be new species that are developing to supplant other species, including ourselves.

STABILITY

Stability implies that there is a dynamic equilibrium between the elements that make up that equilibrium. Stability requires that there are stabilizing networks that direct and process change. For this to happen, the newer, novel elements must not be too destructive to the Self.

In the beginning of our planet, for a period of about a half billion years, Earth's early chemistries eventually began to form membranes that could contain the internalization of other external chemistries. The emergence of these membranes was a Novelty that became crucial for all life. Although, it was the internalization of external chemistry, it also became a place where there developed new chemistries with a complementary relationship between Stability and the Novelty that could be energized by the potential energy gradients across these membranes.

Very early life had to contend with the fragility of its membrane. The first membranes would not have survived in salt water, yet the oceans are believed to be where life began.[1] These issues may be resolved by considering other chemical scenarios.[2] We will continue to experiment, but we may never know exactly what those early chemistries were that created Earth's first life. However, becoming *membrane embedded chemical systems* provided an increased Stability to these chemical systems so that they could weather severe environmental challenges such as drying out or low temperatures. They also served as a source of energy for the first protocells due to the chemical imbalances present within these semipermeable membranes. Therefore, it was vitally important that the correct type of semipermeable membrane could form and maintain itself during protocell evolution. To maintain itself, it may have required a stable, semipermeable membrane template that could form new membranes in an inheritable way.

At some point, with a stable membrane, life had to contend with stabilizing its metabolism. Unrestrained Growth (see chapter 6 for a further discussion) can be destructive to an organism, one example being cancer. Therefore, the earliest life had to develop a way to grow without destroying itself. This early dilemma for life probably involved a long period of time and the interjection of Novelty into an organism's metabolism, which continues to this day. One example of this interjection of Novelty is the discovery of bacteria that eat plastic.[3] Therefore, we know that metabolism continues to evolve and accept Novelty. Once a novel energy source satisfies an organism's metabolic needs, then this incorporation into the chemistry of the cell may stabilize that energy source as an important metabolic resource.

So, too, must the first life have had to contend with these issues of Stability. Life has developed a stable core group of chemistries. These chemistries maintain the integrity of membranes, generate energy gradients

---

1   David Deamer, "The Role of Lipid Membranes in Life's Origin," *Life* 7, no. 1, article 5, (January 17, 2017): 1–17, doi:10.3390/life7010005, https://www.ncbi.nlm.nih.gov/pmc/articles/pmid/28106741/.

2   Koichiro Sadakane and Hideki Seto, "Membrane Formation in Liquids by Adding an Antagonistic Salt," *Frontiers in Physics* 6 (March 27, 2018), https://doi.org/10.3389/fphy.2018.00026.

3   For information on bacteria that can eat plastic, see Darko Petrusheviski, "This Is a Bacteria That Eats Plastic," *Best of Nature*, blog, May 27, 2015, http://optimumnaturae.blogspot.com/2015/05/this-is-bacteria-that-eats-plastic.html. And also Clare Garrard, "Plastic-Eating Bacteria," *Science Accessibly*, blog, April 4, 2016, https://scienceaccessibly.wordpress.com/2016/04/04/plastic-eating-bacteria/.

for use in metabolism, and provide molecules for inheritance. As previously mentioned, heredity isn't included as one of our fundamental criteria for life (see Appendix I B), yet it is one of the most important. We take the position that heredity is an emergent phenomenon. This is a controversial position, but just as other parts of cells have emerged as structural elements, so, too, have our genetic molecules and biochemistries emerged. The early protocells could have relied primarily on a template form of inheritance. This would mean, for instance, that their membranes could form from the ambient molecules available to them much like mineral crystals can form in many environments, given the right conditions. This period may have preceded the catalytic RNA period. It is difficult to appreciate the enormous time spans that certain chemistries may have had to become essential to life. Although much of this is speculation, it provides a sketch of the possible paths taken toward life. Our current genetic heredity is, essentially, the underlying chemistries of proteins, genes, DNA, and RNA. These chemistries arose from the more primitive but adaptive chemistries of the past. The Stability of these core chemistries is essential for the Stability of the Self. These biochemical reactions, by their very nature, encounter and couple with novel but compatible external chemistries.

Stability arises as a balancing of competing chemistries. For every constructive set of chemistries there are a complementary set of destructive chemistries. This balance of competing forces is very much a part of the Essential Tensions, which we discussed in chapter 3. The Stability of these Essential Tensions depends on the structural Stability of membranes and the biochemical support systems that can repair and reproduce these membranes.

Stability also depends on the structural strength of the basic materials used for life forms. Examples include the shells of mollusks and the casings of coral. Some of these structural forms develop in environments that severely test their material strength. The upper limits for how large creatures can be on land may have been tested by some of the titanosaurs.[4] The strength of bones, muscles, tendons, and blood vessels are all critical components that may have neared or reached their structural limits in these gigantic animals during their evolution.

---

4   *Wikipedia*, s.v. "Titanosaur," https://en.wikipedia.org/wiki/Titanosaur.

NOVELTY

There are degrees of Novelty, but for our purpose here, we're referring to Novelty as something new that can also be accommodated by the biochemical or physical responses of a cell or organism. Stability versus Novelty requires a balancing of the metabolic cycles of a cell to produce either useful energy or engage in chemical cycles that are not producing useful energy but could lead to new chemical branches of a Self's metabolism. Life channels energies into chemistry and sets the stage for Novelty. On the most basic level, it does this by having its chemistry poised to connect with other chemistries. These potential positions of life's chemistries also increase the chances for emergence of new properties and chemistries. Because of life's existence, Earth has many different chemistries (and minerals) that it would not otherwise have. Even mankind's pollution of Earth's natural environments holds the potential for the emergence of new, biological, and environmental chemistries.

The metabolic cycle always includes the catabolic use of external nourishment, which also produces waste products as well as chemical reaction products to feed the metabolic cycle and, sometimes, to enable new chemical reactions with external organic compounds. Among the trillions of possible new chemical reactions based on what is available in the environment, a few may lead to new Energy Flows that add to the Energy Flow of the existing metabolic cycle. We call this the search for Novelty, or Neosis: the Self seeks to add components of the external environment to its existing metabolic cycles. We hypothesize that these additions are important parts of Growth in living organisms.

Embryonic development is the most critical time when the environment can shape the development of an organism. Environmental feedback influences the development of many embryos. Examples include crocodile embryos, whose sex is determined by the ambient temperatures to which the eggs are exposed. There have been increases in leukemia among children exposed to x-ray radiation while in their mother's uterus, and diethylstilbestrol (DES) exposure in the womb was linked to reproductive tract abnormalities in the daughters of these women. Some agents produce changes in subsequent generations that have not been directly exposed. There are many other examples of a change in an environmental condition that alters the development of the embryo, demonstrating that environmental changes can directly influence the development of embryos.

Novelty has a dark side where life might exist but isn't quite as successful as its parents. Examples include Thalidomide babies, fetal alcohol syndrome babies and cancers linked to chemical compounds. Life doesn't always handle Novelty well. Poisons fall into this category. It is amazing that such relatively small dosages of poisonous chemicals can have such disastrous consequences for life. Our minds can also be influenced quite surprisingly by a relatively small number of molecules.[5]

THE COMPLEMENTARITY OF STABILITY AND NOVELTY

Once we accept the first fundamental Essential Tensions, it becomes clear that there are many other forms of dynamic tensions that affect both Growth and Stability. While change driven by novel external interactions needs to happen to enable Growth, if external factors drive too much change, then organisms die from insurmountable instability since change without regulatory control is unstable. As previously mentioned, there is a tension between Novelty and Stability and neither can totally dominate the other, but both are in continuous tension with each other.

Novelty doesn't usually threaten destruction; it is an addition to existing structures. It takes what's there and builds on it. Novelty is finding a new chemical pathway that broadens the Self's metabolism. Novelty in a dynamic balance with Stability is what gradually led to the storing of energy and to the maintenance and division of the membrane. It produced metabolism through a gradual accretion of compatible chemistries. It may exist quietly in the background and only be found useful during periods of extreme stress.

Stability versus Novelty defines a dynamic balance between the stability-enforcing powers and molecules of the Self and the novel expansion of potential physical and chemical spaces, such as for new energy sources or chemical functions. There exists a restricted chemical space that defines the overall usefulness of any new proteins or molecules, based on the environmental and

---

5   Michael Polan, *How to Change Your Mind* (New York: Penguin Press, 2018).

biophysical constraints of the Self. This potential chemical space of the Self may be expanded by novel encounters with new exogenous molecules or by creating new proteins/molecules from existing versions. This allows for many important modifications, such as the chemical modifications of existing molecules by many secondary processes, including methylation, phosphorylation, acetylation, and nitrosylation. These chemical modifications regulate almost all aspects of a cell's metabolism and are thought to play important roles in epigenetic regulation, which entails chemical feedback from the environment.

THE ADAPTIVE REGULATION OF STABILITY AND NOVELTY

As life incorporates new Energy Flows into its metabolic cycle, many cells may die before being able to adapt in time to the changes. On the other hand, without some baseline, stable, metabolic cycle, the new Energy Flows alone cannot be self-sustaining. Initially, the balance between how much energy comes from an organism's stable metabolic cycle and what new sources of Energy Flow contribute depends on whatever chemical reactions are embedded within the environment. In this way, the balance between Stability and Growth is self-regulating.

Understanding the effects of chemical interactions on evolution is difficult to visualize because we are not able to see all the feedback loops of the many chemistries being changed by interactions with other external and internal chemistries. The most visible cases are those of bacterial resistance to antibiotics. When we consider the many new chemicals introduced into our own environments from the twentieth century until now, we can hardly begin to understand all these potential interactions and their effects on future generations. This is partly the science of toxicology and public health but is far beyond the capacities of those dedicated scientists to predict the effects of the myriad number of chemicals on future generations. The chemical pollution of our planet will increase both Novelty and Fateful Encounters. Life's present course is sure to change as a result of such new chemistries. We're only beginning to understand how feedback mechanisms operate between a cell's inheritable material and the environment. This is a lively ongoing discussion among many scientists.[6] At its core is the idea that all life must have ways to sense and adapt to different environments. So where does this feedback from the environment to the genome occur? The answer may be that there are multiple ways in which life adapts, including sensing and responding to environmental changes.[7] We argue that this has been a characteristic of life since the beginning and has evolved over these last several billion years to include an impressive array of sensory molecules that can respond to many different changes. The way in which these sensory receptors provide feedback to alter the genome has not been well studied.[8] Functional pseudogenes may be important in controlling genomic drift, which may account for genetic changes that occur when an organism enters a new environment.[9]

We're only now beginning to unlock some of the critical biochemistries that control our cells. Scientists have discovered a gene-editing enzyme labeled CRISPR that may revolutionize how we manipulate the genomes of many organisms, including our own.[10] The point to be made here is that it all boils

6   *Wikipedia*, s.v. "Epigenetics," https://en.wikipedia.org/wiki/Epigenetics.

7   David Julius and Jeremy Nathans, "Signaling by Sensory Receptors," *Cold Spring Harbor Perspectives in Biology* 4, no. 1 (January 2012): 1–14, doi:10.1101/cshperspect.a005991, https://www.ncbi.nlm.nih.gov/pmc/articles/PMC3249628/. And also Joy Alcedo, Wolfgang Maier, and Queelim Ch'ng, "Sensory Influence on Homeostasis and Lifespan: Molecules and Circuits," Madame Curie Bioscience Database, Landes Bioscience (2009), https://www.ncbi.nlm.nih.gov/books/NBK25445/; Sandeep Ravindran, "What Sensory Receptors Do Outside of Sense Organs," *TheScientist*, September 1, 2016, https://www.the-scientist.com/features/what-sensory-receptors-do-outside-of-sense-organs-32942; David Jukam et al., "Opposite Feedbacks in the Hippo Pathway for Growth Control and Neural Fate," *Science* 342, no. 6155 (October 11, 2013), doi:10.1126/science.1238016; Simone Temporal, Kawasi M. Lett, and David J. Schulz, "Activity-Dependent Feedback Regulates Correlated Ion Channel mRNA Levels in Single Identified Motor Neurons," *Current Biology* 24, no. 16 (August 18, 2014): 1899–1904, https://doi.org/10.1016/j.cub.2014.06.067.

8   Jens Rister, Claude Desplan, and Daniel Vasiliauskas, "Establishing and Maintaining Gene Expression Patterns: Insights from Sensory Receptor Patterning," *Development* 140, no. 3 (2013): 493–503, doi:10.1242/dev.079095.

9   Masafumi Nozawa, Yoshihiro Kawahara, and Masatoshi Nei, "Genomic Drift and Copy Number Variation of Sensory Receptor Genes in Humans," *Proceedings of the National Academy of Sciences (PNAS)* 104, no. 51, (December 18, 2007): 20421–20426, doi:10.1073/pnas.0709956104. And also *Wikipedia*, s.v. "Pseudogene," https://en.wikipedia.org/wiki/Pseudogene.

10 *Wikipedia*, s.v. "CRISPR," https://en.wikipedia.org/wiki/CRISPR.

down to an understanding of chemistry. This suggests that we may be missing some novel and vital part of chemistry that is important for life. All components of existing life interact chemically with the environment and internally with other molecules including genes. Life is an evolving set of chemistries. As Michael has argued, Quintessence applies to all life forms and in the process revises the story of evolution and revises our own self-image. These processes are inherited from chemistry when life first formed and operate at all levels of life from bacteria to plants to animals to humans to societies. Some examples of Quintessence in action include how DNA originated when life first formed, the evolution of simple cells to nucleated cells, how multicellular organisms began, and how genetic Novelty leads to the evolution of species. There are specific examples of Quintessence that include how bacteria quickly gain resistance to antibiotics,[11] how the CRISPR-Cas system defends bacteria from viral invaders,[12] the discovery of bacteria that eat plastic, and how temperature changes alter the development of embryos.[13]

METABOLIC PATHWAYS

We will make progress using our increasing computational power with artificial intelligence (AI) to assist us in following each chemical pathway and determining how it interacts with the myriad of other chemical pathways within living cells. We may often lose sight of the bigger picture, which is to put the parts into the context of the whole organism. Hopefully, this book will help to maintain a suitable perspective. It is miraculous that a relatively small set of a dozen or so of the basic chemical elements could come together in this chaotic universe for billions of years to maintain something we call life!

---

11 Eugene V. Koonin, and Yuri I. Wolf, "Is Evolution Darwinian or/and Lamarckian?" *Biology Direct* 4, no. 1 (November 2009): 42, doi:10.1186/1745-6150-4-42.

12 X. Feng, and S. Guang, "Small Rnas, Rnai and the Inheritance of Gene Silencing in Caenorhabditis Elegans," *Journal of Genetics and Genomics* 40, no. 4, (April 20, 2013): 153–160, doi: 10.1016/j.jgg.2012.12.007, https://www.ncbi.nlm.nih.gov/pubmed/23618398. And also Sophie Juliane Veigle, "Use/Disuse Paradigms Are Ubiquitous Concepts in Characterizing the Process of Inheritance," *RNA Biology* 14, no. 12 (December 2, 2017):1700–1704, https://www.ncbi.nlm.nih.gov/pubmed/28816621.

13 *Wikipedia*, s.v. "Temperature-Dependent Sex Determination," https://en.wikipedia.org/wiki/Temperature-dependent_sex_determination.

The reconstruction of the history of life on Earth represents one of the most intriguing issues of science. And even more intriguing is trying to understand the very first molecular steps leading to the primordial cells and their early evolution. Cells are quite complex entities constituted from a myriad of different molecules that act and interact in a concerted manner to assure the survival and reproduction of cells and multicellular organisms. In each moment of a cell's life, billions of molecules are transformed into different ones through reactions that are catalyzed by enzymes (proteins). Even though these enzymes might interact with a plethora of different molecules during their chaotic trip within the cell, they also operate on specific molecules called *substrates*, and thereby transform these *substrates* into different molecules called *products*. Hence, in each moment of a cell's life, *billions* of substrates are transformed into *billions* of products by *billions* of enzyme molecules. Together, these reactions constitute *metabolism*, a circular entity in the sense that molecules can be either destroyed or catabolized to obtain energy, or they may be created by a process called anabolism. It is thus clear that, within a cell, an equilibrium between catabolic and anabolic reactions exists. Feeding the existing metabolic cycle and adding new internal or external Energy Flows play complementary roles in the survival of life. Thus, the metabolism of living cells is quite complex, but we can also consider it extremely ordered.

The term *metabolism* describes the web of chemical reactions that maintain the living state of cells and organisms. This includes both reactions that synthesize amino acids and lipids that cells need and reactions that break down molecules to generate energy. Cells use lipids and amino acids in membranes and proteins and to create the molecules that are consumed to generate energy. A metabolic pathway is a linked series of chemical reactions occurring within a cell. In most cases of a metabolic pathway, the product of one enzyme acts as the substrate for the next. Different metabolic pathways function based on their position within a cell and the significance of the pathway. For instance, the citric acid cycle, electron transport chain, and oxidative phosphorylation all take place in the mitochondrial membrane. In contrast, glycolysis, the pentose phosphate pathway, and fatty acid biosynthesis all occur in the cytosol of a cell.

In addition to the two distinct metabolic pathways of anabolism and catabolism, there is the amphibolic pathway which can be either catabolic or anabolic, based on the need for, or the availability of, energy. Pathways are required for the maintenance of homeostasis within an organism, and the flux of metabolites through a pathway is regulated, depending on the needs of the cell and the availability of the substrate.

In my conversations with Michael, he strongly felt that one should realize that the general properties of homeostasis (Stability) and change (Novelty) are necessary to understand both Growth and evolution. What are the important properties of Novelty and Stability that affect Growth and evolution? Homeostasis, and by implication, Stability, is about an organism's self-regulation to maintain survival with its current set of properties. If homeostasis ruled all organisms, nothing would grow or evolve, and soon enough, that totally homeostatic organism would die from competition, or because its metabolism became dysfunctional in the context of new environmental circumstances.

Any permanent point change to the organism must be allowed by homeostasis and must benefit the organism. Benefit is defined here as improving the energetics of metabolism either by efficiency or by increasing the available energy that can be processed by the organism. Permanent change to homeostasis occurs either by regulatory change above a threshold that becomes permanent, such as a change to the regulatory mechanism itself, or by a chance encounter from external interactions that fits a cell's metabolism by increasing the energetics of the internal metabolism. The nature of this fit is based on chemical reaction properties and is the most fundamental mechanism of choice at the cellular level.

CHAPTER 6

Growth and Energy Flow

"The key to growth is the introduction of higher dimensions of consciousness into our awareness."

—Lao Tzu

W e live lives of Growth. Usually we think of Growth in terms of adding more to what we already have, such as weight, wealth, power, status, and so on. We all grew from our mother's egg, which was about the size of the period at the end of this sentence, to our adult size of about twenty billion times larger than the original size of the fertilized egg.[1] We take for granted this phenomenal amount of Growth![2]

---

1 *Wikipedia*, s.v. "Orders of magnitude," The human ovum weighs about 3.6 x 10^{-9} kg and the average human weighs about 70 kg, so 70/3.6 x 10^{-9} = 19.44...x10^{9}, https://en.wikipedia.org/wiki/Orders_of_magnitude_(mass).

2 You began as your mother's egg cell that was fertilized by your dad's sperm. This began a miraculous series of events leading to the morula stage. The morula is produced by a series of cleavage divisions of the early embryo, starting with the single-celled zygote. Once the embryo has divided into sixteen cells, it begins to resemble a mulberry, hence the name *morula* (Latin, *morus*: mulberry). Within a few days after fertilization, cells on the outer part of the morula become bound tightly together. This process creates a cavity inside the morula, which results in a hollow ball of cells known as the blastocyst. The blastocyst's outer cells will become the first embryonic epithelium (the trophectoderm), which will form the placenta. Some cells divide into the interior and become the inner cell mass. They are

When we were born we weighed on average about 3.5 kg (7.7 lbs; the normal range is about 6.5–8.0 lbs., but this very much depends on the group being sampled) and we had about 3.5 x 10^{12} cells (with an average cell weight of about one nanogram per cell, 1 x 10^{-9} g), which suggests that from that first fertilized egg cell, there were about forty-two cell divisions necessary to form our bodies as a newborn baby. You may wonder how only forty-two cell divisions create a trillion cells. Well, calculating 2^{42} is 4 trillion, 398 billion, 46 million, 511 thousand, and 104. Doubling repeatedly is a powerful method for increasing size relatively quickly. We humans grow in our societies and through education. We grow our spheres of influence over possessions, people, money, and power. Our horizons expand in a widening spiral. We emerge from the microbiological world into the magical realm of other beings, all while attempting to grow!

Nothing grows without the input and transformation of energy and resources. The handling and manipulation of Energy Flows by the Self leads to Growth. Growth includes the storing and copying of the essential parts of the Self (heredity), as well as energy storage in molecules, or as the Essential Tensions in membrane gradients. Growth depends on the ability of life to harness these Energy Flows. This suggests that Growth needs both an energy source and the physical structures that can use these energy sources. For our cities, we also need suitable structures that harness energy sources such as nuclear reactors, solar panels, wind turbines, hydro-electric generators, and so on. This demonstrates the critical need to maintain those suitable structures that can sustain the necessary Energy Flows, whether for a city or a living organism.

To be successful, a Self must manipulate Growth and Energy Flow into productive lines. Growth is the process whereby life captures Energy Flows and directs them to creating more components of the Self. This is accomplished by the manipulation of matter to form parts of the Self. Energy Flows allow the

---

pluripotent and, ultimately, form the embryo. That's how the authors and readers of this book got their start—a rather complex process that no one could have thought up before! This example illustrates the myriad complexities of just a small fraction of the biological processes that we take for granted everyday. If nothing else, this book should show just how much more we need to learn.

chemical reactions of the cell to combine atoms from the chemical elements and other more basic molecules into larger and more complex molecules (a molecule consists of two or more atoms and an input of energy is usually required to form one). This process of combining atoms or molecules is done repeatedly to grow the cell or organism.

According to the theory of dissipative structures, an open system has the capability to continuously import free energy from the environment and, at the same time, export entropy. Growth and Energy Flows are intimately connected. In general, Energy Flows are directed by the molecular structures that collect the energy to be used by other molecules that store the energy or that use the energy to build more structures within the Self. These Energy Flows must be in a range of magnitude that life's molecules can use. If the Energy Flows are too large, life may be destroyed, and if they're too small, life may not be able to utilize the available energy. Using some portion of the Energy Flow to copy Self also requires energy. Excess Energy Flows are either dissipated or stored in some molecular form, such as ionic gradients across a membrane (the Essential Tensions).

Growth and Energy Flows are connected by the metabolism occurring within the Self. That is not to say that all metabolism must occur within the Self. There may be important products or cofactors produced by other organisms or chemical processes within the environment that are necessary for the metabolic cycles within the Self. This overall metabolism has cyclical periods of increase and decrease. These changing environments may slow the Self's metabolism, which creates the necessary condition for dormancy, or stasis of the Self. This is observable in all forms of life. Bacteria have spores; plants have seeds; animals sleep or hibernate, and so on.

The original Self must have possessed some ability to remain dormant to survive in hostile and variable environments. Even supposing that the original Self began at a relatively constant hydrothermal vent, the act of leaving that environment must have necessitated the need for a dormant state since it could no longer depend on its relatively constant food supply (or Energy Flows) at the hydrothermal vent. Today it is becoming more certain that all life has this innate ability to some degree. This also raises the issue of Energy Flow and thermodynamics in ancient Selves. Obviously, if the first primitive

Self was in a dormant state, it was also in thermodynamic equilibrium with its surroundings. Spores and seeds last for years or decades in this dormant state, only waiting for the right environmental conditions, such as water or fertilization, to begin their metabolic machinery again. -So too must it have been for the original Self. This means that Growth can be put on hold, which is an overlooked point in most evolutionary theories. We will discuss dormancy in more detail later in this chapter.

Unrestrained Growth is destructive. Cancer is an example of this. The earliest life had to develop a way to grow without destroying itself. This became an early dilemma for life. One problem is that as a cell grows larger, its surface area to volume ratio decreases by the ratio of $3/r$, where r is the growing radius of the cell. As the radius increases, the surface area to volume ratio decreases. This means that the surface area to volume ratio of growing cells gets smaller as each cell grows. A smaller surface area relative to the cell's volume means that less nutrients can diffuse into the interior of the cell. This was a significant problem for the simpler cells of earliest life because they depended on the ability of essential nutrients and other molecules to diffuse into the interior cytoplasm of the cell. If this didn't occur, the metabolic machinery of these simpler cells would have been disrupted. Egg cells avoid this problem by having their essential nutrients already stored within their cytoplasm. The more modern eukaryotes solve this problem by having various forms of active transport within their cytoplasm, but the earliest life must have relied only on diffusion. This established a limit to the size a cell could grow before it had to do something, such as divide or elongate. But what was it that limited the size of these earliest cells?

What might have the protocells done before life? Can protocells grow? The short answer is that they can grow, given the right environmental conditions, which is also true for living cells. Protocells often need to have other, external physical forces present to force them to divide into smaller cells, unlike living cells, which appear to provide their own forces for cell division. What might have been the first and most primitive mechanism for cell division in the earliest cells?

Cells eventually controlled their reproduction by linking their metabolism to the process of cell division. This became a significant emergent leap for cellular evolution and involves a complex dance with the many cellular molecules and enzymes that control Growth. We're only beginning to unravel these processes; we still have much more to learn about how the coordination is achieved. One promising area of research is the cell membrane since it is required to increase its area large enough to cover two daughter cells, which form after the original mother cell divides. Together, these daughter cells have a larger surface area than the mother cell although they have the same volume (see Appendix V). This is because it takes more surface area to cover a volume that is divided into smaller volumes.

In the intriguing book *Scale* by Geoffrey West, we're introduced to the universality of a Growth law for all life.[3] West uses concepts of physics to explain how life, cities, and companies operate on an underlying law of scale. One important concept connecting life, cities, and companies is that they all consume some type of energy, or money as a form of energy, and they all exist as connected networks. They all have a metabolism that drives their operation. The fact that there are similar scaling laws for these different entities leads one to question why. One of the most important constraints to Growth is diffusion, which West mentions briefly. The diffusion through networks that are essential for Growth may be what primarily determines this scaling law.[4] Tortuosity (many turns or twists) can greatly influence the rates of diffusion through networks. In general, the greater the tortuosity, the slower the diffusion. These are important physical restraints on living organisms and cities. Diffusion also works in both directions and must therefore be balanced. It transports the molecules necessary for metabolism and removes those that are waste.

ENERGY FLOWS

Capturing Energy Flows is perhaps one of the most important functions of living organisms. What this means in practice is that the organism taps into an Energy Flow in its environment. We're just beginning to understand that almost any energy gradient can be tapped by living organisms. The sun's energy

3   Geoffrey West, *Scale: The Universal Laws of life, Growth, and Death in Organisms, Cities, and Companies* (New York: Penguin Books, 2018).

4   Mathematically consistent constraints are created by the tortuosity of networks in conjunction with diffusion. Tortuosity is commonly used to describe diffusion in porous media. See *Wikipedia*, s.v. "Tortuosity," https://en.wikipedia.org/wiki/Tortuosity.

is thought to provide most of life's energy on Earth, primarily through plants' photosynthesis, but there is also a substantial percentage of energy that is supplied by geochemical or geothermal energy gradients.

The evolution of life depends on the initial conditions of the first environments and, subsequently, the incremental Growth produced by the novel creation of Energy Flows. Most beginnings of life die before gaining any foothold, and those that make it, by our logic of Growth, explode in diversity. The Growth and evolution processes become exponentially more complex as multiple living organisms begin to interact with each other in both mutually beneficial and detrimental ways.

It is interesting to note that most of the current hypotheses concerning the origin of life don't cite sunlight as the energy source for beginning life. Geothermal and geochemical sources are often cited as the probable energy sources for the earliest life. This raises our hopes for discovering life on other planets farther from the sun.

Generally, cellular respiration is categorized as either aerobic or anaerobic, whereas fermentation is an additional modification to low oxygen levels.[5] These processes have evolved over billions of years of evolution. How the cell's metabolic machinery breaks down Energy Flows is a vitally important process. During the four billion or so years of evolution, new Energy Flows have emerged. Life has developed ways to store excess energy into metabolic reservoirs. These are also emergent phenomena that have served life well in times of disrupted Energy Flow. Excess Energy Flows must be either dissipated or stored in some molecular form—for example, adenosine tri-phosphate (ATP)—or as an Essential Tension, such as ionic gradients across a membrane, or in other complex molecules, such as lipids or fats.

Microbes can eat many things.[6] Bacteria living in hot springs or without

---

5 *Wikipedia*, s.v. "Cellular respiration," https://en.wikipedia.org/wiki/Cellular_respiration. See also *Wikipedia*, s.v. "Anaerobic respiration." Note the very nice table listing the various types of respiration organisms use and the products or waste from these processes, https://en.wikipedia.org/wiki/Anaerobic_respiration.

6 Ken Nealson, "Extracellular Electron Transport (EET): Opening New Windows of Metabolic Opportunity for Microbes," Wiley Environmental Microbiology Annual Lecture 2015, YouTube video, Nov 26, 2015, https://www.youtube.com/watch?reload=9&v=qBOWuMz-RaU.

oxygen are impressive, but then there are also bacteria that live off the energy from pure electricity, feeding directly on electrons![7] Even more surprisingly, scientists find that these bacteria are not even that rare. "Stick an electrode in the ground, pump electrons down it, and they will come," begins a fascinating *New Scientist* piece on these electricity-eating microbes. For a while now, we've known about two types of bacteria that eat electricity, *Shewanella* and *Geobacter*. As scientists stick their electrodes in wells, marine mud, and gold mines, they've found several more types of electricity-eating bacteria. These bacteria are getting energy in its purest form as electrons. The rest of *all* known life on Earth also moves electrons around for energy, but the process is managed through a complex set of reactions involving ubiquitous energy molecules such as ATP. This begs the question of what molecules these electrons are moving onto to create the energy and matter necessary for Growth. There is always an underlying structure that takes in the Energy Flow—or in this case, electrons—and channels it to create other molecular structures that store this energy within the cytoplasm of the cell.

We argue that the free energy conversion mechanisms that began at life's emergence were the disequilibria (Essential Tensions) that formed during the development of the semipermeable membranes of the progenitor protocells. These free energy-conversion mechanisms were not, by themselves, sufficient for life, because they lacked the sensory feedback mechanisms to sense and adapt to the environment. It was the coupling of sensory feedback with the first primitive Energy Flows that led, eventually, to life.

Chemically, the Energy Flows resulting from redox disequilibria act like a battery for life.[8] The term *redox disequilibria* is just a fancy way to say that a battery can form in nature when there is something in between that can conduct electrons. As mentioned before, there are bacteria that can eat electricity. They

---

7 Catherine Brahic, "Meet the Electric Life Forms That Live on Pure Energy," *NewScientist*, special report, July 16, 2014, https://www.newscientist.com/article/dn25894-meet-the-electric-life-forms-that-live-on-pure-energy/. And also Emily Singer, "New Life Found That Lives Off Electricity," *Quanta Magazine*, June 21, 2016, https://www.quantamagazine.org/electron-eating-microbes-found-in-odd-places-20160621/#.

8 Wolfgang Nitschke, "Life as a Dissipative Structure," presentation at Conference, "From Star and Planet Formation to Early Life," Vilnius University, Vilnius, Lithuania, April 25–28, 2016, YouTube video, https://www.youtube.com/watch?reload=9&v=duPNezyS1gg.

have the metabolic machinery that is necessary to change the electric energy they feed on to create new and more complex molecular configurations, or to store the energy as either molecular or membrane gradient potential energies.

When we consider the cost benefit analysis of various sources of energy in society such as oil, gas, coal, nuclear, wind, solar, or hydroelectric systems, we recognize that each energy production system requires specific structural mechanisms that capture, transform, or produce energy from a specific source. The true cost-benefit analysis of each of these systems is more complex than it initially appears since we should also consider the costs of building and maintaining the specific structures for capturing, utilizing, and distributing these energies. The same is true for life.

The metabolic cycle always involves the catabolic use of external nourishment, which includes waste products as well as chemical reaction products that both feed the metabolic cycle and enable new chemical reactions.[9] Among the trillions of possible new chemical reactions, based on what is available in the environment that might permeate the cell wall, a few lead to new Energy Flows that add to the existing Energy Flow of the metabolic cycle. We call this the Fateful Encounter between the Self and its environment. Such an encounter stimulates the Self to add components of the external environment to its metabolic cycle. We hypothesize that these additions are the definition of Growth in living organisms, which includes the incorporation of photosynthesis as the most powerful new source of Energy Flow initially developed in *archea*.[10]

RECRUITMENT AND SYMBIOSIS

The cooperation of organisms may lead to improved Energy Flows as well as other benefits. Cells in colonies will chemically attract and repel each other through chemical communication from escaped products due to osmotic

---

9   According to the theory of dissipative structures, an open system is capable of continuously importing free energy from the environment and, at the same time, exporting entropy.

10   Martin F. Hohmann-Marriott and Robert E. Blankenship, "Evolution of Photosynthesis," *Annual Review of Plant Biology* 62 (June 2011): 515–548, https://www.annualreviews.org/doi/abs/10.1146/annurev-arplant-042110-103811.

gradients in the semipermeable, outer-cell membrane. This creates the emergence of colony interactions that can lead to group synchrony, or recruitment, or subgroup symbiosis of chemical reactions that may provide benefits of an increase in Energy Flow to the cells or organisms involved.

SELF-BENEFIT AND SELF-REGULATION

We debated the possible mechanisms that could explain Self-benefit. We came to the conclusion that the underlying basis of Self-benefit is the organic chemical reactions that proceed from reactants to products. The signal for Self-benefit is the reaction itself and the complement, which is the cessation of existing chemical reactions, is the signal for Self-detriment, degradation, dormancy or death.

Life can only continue to grow if the Energy Flow needed to continue the metabolic cycle is controlled and directed through an autocatalytic process. An essential criterion for this is the presence of a chemical feedback loop that can sense and respond to changes in the Self's Growth and Energy Flow. In the first metabolic cells, reactants in the metabolic cycle needed immediate sources of Energy Flow to maintain their autocatalytic chemical reactions. Later, the creation of energy storage with molecules such as ATP allowed the transfer of Energy Flow to the reactions essential for the maintenance of the Self to become more regulated and even. This was another of many emergent phenomena that led to the evolution of modern cells.

We revisit homeostasis here to put it in the context of our discussions concerning Growth and Energy Flow and Self-benefit and Self-regulation. Any permanent change to the organism needs to be allowed by the overall homeostasis and be a net benefit to the organism. Benefit is defined here as improving the energetics of metabolism either by efficiency or by increasing the available energy that can be processed by the organism. Permanent change to homeostasis can occur through regulatory change above a threshold that becomes permanent, such as a change to the regulatory mechanism itself. It can also be produced from outside interactions by a chance encounter, which is internalized through an immediate increase in the energetics of the internal metabolism.

All homeostatic control mechanisms have at least one sensing receptor that monitors and responds to changes in the environment, either external or internal. Receptors include thermoreceptors, mechanoreceptors, and nuclear receptors that bring about changes in gene expression through up-regulation or down-regulation of gene expression.[11] All together these act as feedback mechanisms to promote homeostasis. There is still much to learn about how these receptors determine the feedback for homeostasis and the expression of genes for embryo development. It is important to emphasize that homeostatic reactions are maintained by many systems operating together.

Michael's comments regarding "the most fundamental mechanism of choice" being "based on chemical reaction properties" is a major concept in this book that I've developed beyond our previous conversations. I maintain that the net shifts that may occur within the Essential Tensions are these chemical reaction properties that lead to the most fundamental mechanisms of choice. When Essential Tensions become linked together as networks, the whole becomes much greater than the sum of its parts. The importance of these biological/chemical systems in the trajectory of life has been previously overlooked, which is perhaps due to the perceived complexity underlying these systems.

FEEDBACK

Feedback is the primary way that life controls Growth and Energy Flows into its metabolic cycles. Feedback, an important concept for understanding life, is generally thought to occur in two ways. It can be positive, which promotes the process, or negative, which hinders the process. However, there are multiple ways in which both positive and negative feedbacks can produce modulation in key centers of cellular metabolism.

The term *cybernetics* is applied to technological control systems such as thermostats, which function as homeostatic mechanisms. Cybernetic systems, whether in the machine or in the animal, can remember disturbances and thus are used in computer science to store and transmit information.

---

11 *Wikipedia*, s.v. "Nuclear receptor," https://en.wikipedia.org/wiki/Nuclear_receptor.

Negative feedback is a central homeostatic and cybernetic concept, referring to how an organism or system automatically opposes any change imposed upon it. Positive feedback can be more destructive than negative feedback. Disturbance, or departure from equilibrium, is every bit as important as negative feedback: Systems cannot correct themselves if they do not stray.

Basically, even simple, two-state receptor, or enzyme, systems can be modulated (see Appendix III). This negative feedback is automatically encoded in the biophysical mechanisms of these two-state systems. One important example of this is the responses of our senses through receptor molecules. When there is too much stimulation, our senses become dulled (other terms to describe this process include desensitization, tachyphylaxis, down-regulation, habituation, autoinhibition, and tolerance).

Most enzyme systems are similar to these two-state sensory systems, which is precisely how systems of negative feedback have developed intrinsic to these specific biochemistries (as autoinhibition or substrate inhibition in the case of enzymes). The dualism concept of balance suggests that our sensory receptors will either recognize a signal or not. The modulation of this simple on-or-off switch contains multiple complexities when considered in the larger context of the dose-response relationships of our molecular, sensory receptors.

Oscillation is a common and necessary behavior of many systems. In other words, each feedback is less than the last departure from the goal, so the oscillations dampen. Feedback takes time and such a time lag is an essential feature of many natural systems. This causes systems to oscillate above and below their normal equilibrium set point.

WASTE

Humans have become one of the largest and most varied waste producers on Earth, but waste is a relative term. Wastes are the end-products of an organism's metabolism. Both catabolism and anabolism produce waste products that organisms need to remove from themselves and their immediate environment. One organism's waste often is another organism's food or essential requirement. For example, plants produce oxygen as a waste product of their metabolism, which is essential to animals. Animals produce carbon dioxide as a waste product, which is essential to plants.[12] Removal of waste becomes a factor when it creates feedback that affects the Growth of the Self. Life has evolved to fill almost every niche on Earth and in doing so, it has established an overall balance crucial to which is its ability to respond to the opportunities presented by the wastes of other life forms.

HEALTH

Health is a concept that needs to be clarified in terms of the Growth and Energy Flows available to any organism. Just as a person's overall health can be ascertained by the number of options they have, so too can the overall health of any living organism be determined by the number of options it has. For people, this applies to our social, psychological, and financial health. For organisms, this applies to their social networks among species as well as their connections to other environments. Some of these connections can be mutually advantageous. Connecting with and being supported by connections with other species promotes options for optimal health but may also contribute to new diseases. Once again, these processes represent the dynamic balance critical for the fundamental processes of life and evolution.

THE EGG AND THE ZYGOTE

The emergence of the zygote came much later in evolution, sometime after the eukaryotes emerged about two billion years ago. The egg cell is the largest single cell that we can see. Once an egg cell is fertilized by the sperm, it becomes a zygote. This is where we all originated. Since a zygote is a fertilized egg cell, it contains everything necessary to form a new Self.

Ontogeny is the development processes an animal undergoes from egg to adult. The old saying is that ontogeny recapitulates phylogeny, which means that during the development of the embryo, the organism passes through the previous stages of evolution that contributed to the creation of the current organism.

---

12 American Chemical Society, "Joseph Priestley and the Discovery of Oxygen," https://www.acs.org/content/acs/en/education/whatischemistry/landmarks/josephpriestleyoxygen.html.

Sexual reproduction developed before the zygote. Once the earliest life formed, which is estimated to be about four billion years ago, life may have existed as prokaryote and archaea cells for the first 2.5 billion years or so of life's early history. Prokaryotes have a primitive form of sexual reproduction called conjugation, which is the transfer of genetic material between cells by direct cell-to-cell contact or by a bridge-like connection between two cells. In the eukaryotes, all the multicellular plants, animals, and fungi reproduce sexually. It isn't known exactly when sexual reproduction first appeared, but some date it to about 1.2 billion years ago in the Proterozoic Eon. Sexual reproduction does occur in single-celled protists, which are eukaryotes. They include protozoa (unicellular and animal-like, such as amoebas), protophyta (plant-like, such as single-cell algae), and molds (fungus-like, such as slime molds).

In the case for the gentle and kind reader of this book, the formation of your "Self" began as an egg cell from your mother.[13] In addition to receiving your mother's DNA from her genes, all your mitochondria, original cell cytoplasm, RNAs and proteins with post-translational modifications came from your mother's egg cell. That's much more of an inheritance than you may have understood. Your mother's egg was a poised catalytic system waiting for that first penetration through its membrane by the head of a sperm cell containing not much more than the DNA and some RNA from the father.

Pause here and think how much work lies ahead for this now-fertilized mother cell! First, the fertilized mother cell has to activate the cellular machinery that's been dormant for a relatively long time, copy the father's DNA from the sperm and copy the mother cell's DNA to what is soon to become a very large number of new daughter cells. Second, she's got to gear up her membrane-making machinery because each of the many new daughter cells require a new membrane to help cover them (see Appendix V). Third, she's got three to five days to float through the fallopian tubes and then find and attach herself to a spot in the uterus.

---

13 *Wikipedia*, s.v. "Embryonic development," https://en.wikipedia.org/wiki/Embryonic_development.

EMBRYO DEVELOPMENT

Our development in the womb was a series of complex events that were also susceptible to external environmental factors. The nuclear receptors orchestrate the on-and-off activation of genes in the developing embryo. Although the nuclear receptors have been recognized at least since the 1980s, the roles that they play in determining the activation of genes requires much more research.

Miscarriages in humans are surprisingly high. Estimates range from 5 to 25 percent, or more in some cases. We're usually not consciously aware of how fortunate we are to have made it to birth. But why should birth be this costly? The miscarriage rate in animals may be lower than 5 percent, but why would life evolve into this set of circumstances? The reasons are many, but they represent just another mechanism of selection that is a necessary mechanism of biology and evolution. It isn't about what's most efficient for the individual organism.

DORMANCY

Nature is unpredictable. Temperature swings, availability of Energy Flow, and stresses are a few of the factors that can change the environment of a living organism and greatly diminish or stop its Growth. Many microorganisms respond to variable environmental conditions by entering a reversible state of reduced metabolic activity, known as dormancy. Some living microorganisms thousands, hundreds of thousands or even millions of years old have been retrieved from ancient materials.[14]

Microorganisms have evolved a diverse set of mechanisms that allow individuals to enter and exit a dormant state. These mechanisms include but are not limited to the ability of cells to regulate cellular metabolism, form long-lived endospores, enter a viable but non-culturable state, and produce protective resting stages that are formed during asexual or sexual reproduction. Dormancy has attracted attention in the clinical realm because it can help explain how pathogens tolerate high concentrations of antibiotics.

---

14 Jasmin Fox-Skelly, "Some Lifeforms May Have Been Alive Since the Dinosaur Era," BBC Earth, June 3, 2016, http://www.bbc.com/earth/story/20160602-some-lifeforms-may-have-been-alive-since-the-dinosaur-era.

Microorganisms can transition between active and dormant states. Dormancy has important consequences for evolution because it can provide a fitness benefit for organisms in hostile environments. Dormancy is often regulated by environmental cues, such as changes in temperature, pH, water, and resource supply. The fact that viable microorganisms have been retrieved from materials that, in some cases, are hundreds of millions of years old suggests that dormant microorganisms can survive for a far longer time than actively-reproducing individuals. Because large numbers of various microorganisms can enter a dormant state, it is likely that dormancy has influenced the evolutionary history of many microorganisms.

The ecological environment has varying periods of plenty followed by scarcity, when the Self's metabolism slows enough to enter some form of dormancy (or stasis). This is observable in all forms of life. Bacteria have spores, plants have seeds, animals need to sleep or hibernate, and so on. The original Self must have had the original requirement of dormancy to survive extreme conditions in variable environments. Even supposing that the original Self began life at the environmental conditions of a relatively constant hydrothermal vent, the act of leaving that environment must have necessitated the need for a dormant state. This also raises the past issues of Energy Flow and the overall thermodynamics of life. Obviously if the first primitive Self was in a dormant state, it was also in thermodynamic equilibrium with its surroundings. Spores and seeds last for years in these dormant states, only waiting for the right environmental conditions such as water to begin their metabolic machinery again. This must have been also true for the original Self when life began. This means that Growth being put on hold is an overlooked fact in many evolutionary theories.

In many cases, without new Growth, new metabolic cycles may become disrupted and dormant for extended periods of time. The Self is only in an evolutionary context when compared to a surrounding environment, but it may separate itself from this environment. This is not to say that the Self does not depend on the environment, although there may be times that the Self lies dormant or in a state of near-dormancy. The Self can be separated from its environment for varying periods of time without harm, which may occasionally involve a form of dormancy or stasis. Dormancy may also allow much longer exposures to relatively harmful environmental conditions, such as radiation or toxins. Dormancy sometimes comes at a cost because dormant organisms don't reproduce and some organisms must invest endogenous resources into resting structures and maintenance of minimal energy requirements.

In higher forms of life, dormancy had to evolve into a much more complicated process than that practiced by the earliest life forms. Particularly for the eukaryotes with their intracellular organelles (such as mitochondria and chloroplasts), the dormant state had to include some maintenance of these intracellular organelles, which entails dealing with their minimal energy requirements and removal of any waste products.

Dormancy is an adaptive trait that has independently evolved multiple times across the tree of life. By entering a dormant state, individuals can endure conditions that are suboptimal for Growth and reproduction, thereby increasing a population's long-term fitness. We're trying to understand the molecular mechanisms for dormancy, which would benefit those attempting space travel as well as those who require anesthesia during surgery. Related to dormancy is the phenomenon of suspended animation, which has recently received attention.[15] These are fascinating subjects, which science is just beginning to explore.

---

15 Mark Roth, "Suspended Animation Is within Our Grasp," TED Talk, 2010, https://www.ted.com/talks/ mark_roth_suspended_animation/discussion?language=en. And also *Wikipedia*, s.v. "Suspended animation," https://en.wikipedia.org/wiki/Suspended_animation.

CHAPTER 7

Emergence

"Only through art can we emerge from ourselves and know what another person sees."

—Marcel Proust

WHAT IS EMERGENCE AND WHAT HAS EMERGED?

Emergence is the beginning of something new that is scientifically unexplainable. Based upon our current knowledge, we see no logical path to connect all the dots between previous examples and the new phenomenon. For example, the emergence of a butterfly from its chrysalis can be analyzed from multiple perspectives. It is truly an emergent phenomenon when viewed from the perspective of evolution. When we humans began to control fire, we couldn't have predicted the gasoline engine. When Ben Franklin found that he could trap lightning by flying his kite, he couldn't have predicted the electric light bulb or computers. For those with a more religious bent, emergence is a source of the miraculous. On an artistic and cultural level, who could have predicted the Sistine Chapel or the Parthenon? When life first began, we couldn't have predicted that you would be reading this book!

Emergence is one of the great mysteries of life. All forms of life, including the Essential Tensions, are emergent phenomena. Quintessence may provide an explanation for how this might happen, as life's innate ability to internalize Novelty. Michael's goal was to expand the current views of life and evolution by suggesting that being alive creates new opportunities for many emergent phenomena. This expanded view is an important theme of this book. As we learn more about the science of life, how we cope with this knowledge will become a vitally important issue that Michael wanted to address.

For all our scientific progress, we frequently can't describe or understand an emergent phenomenon. Emergence itself is a new area of serious study,[1] in which we often attribute the law of unintended consequences. Without having complete knowledge of any situation, we may be surprised by real-life outcomes. Since no person is all-knowing, most of us can expect to be surprised by the emergent properties of many things that we think we know quite well but really don't. Emergence arises from the many properties that are sources of change common to all life. These sources of change include stresses, such as pain, and new circumstances resulting from Essential Tensions, Novelty, and Fateful Encounters. These are basic properties that we don't often consider but that often direct our lives.

One of the most basic emergent phenomena is the concept of a simple equilibrium. Since the beginning of the universe, it appears that physical evolution came first, followed by chemical evolution, and eventually, biological evolution. Early in this period of the universe's evolution, physical and chemical processes began engaging in back-and-forth dynamic exchanges that created what we label *equilibria*. These fundamental processes began to balance each other and engage with other forces that eventually formed new tensions from these back-and-forth engagements, together with other processes that interfered with them. These, eventually, created the equilibria that led to life's essential disequilibria. Within this book, we discuss nine areas of emergence that have played important roles in the emergence of life and evolution:

1. Emergence of chemistry
2. Emergence of metabolism
3. Emergence of Self
4. Emergence of life
5. Emergence of information

---

1 *Wikipedia*, s.v. "Emergence," https://en.wikipedia.org/wiki/Emergence.

6. Emergence of heredity
7. Emergence of evolution
8. Emergence of species
9. Emergence of complexity

These nine areas don't specify what had to have happened first, second, third, and so on, but they focus our attention on specific areas that are considered important by many scientists who've studied the processes considered to be essential for life and evolution.

In a very real sense, emergence is life's magic. We should not fear the use of the word *magic* in a scientific context, because so many wonderful concepts and ideas were magical when they were first explored by generations of past scientists. The ancient alchemists searched for the distinction between the real and the magical, which led to the science of chemistry. We haven't calculated everything and we're far from figuring it all out. The magical is built into life's core, which is a great gift that keeps us guessing and searching for the bedrock principles underlying our quest for knowing and understanding the universe. We are beings who must contemplate and embrace this mysterious, emergent, and magical universe because it is also inherent within each one of us.

EMERGENCE OF CHEMISTRY

We've previously mentioned that the reader is an amazing being who represents the culmination of billions of years of evolution. As the universe expanded and stars formed, the chemical elements became much more complex. The evolution of chemistry exploded as millions upon millions more chemical compounds formed from the universe's early chemical elements. These chemistries further evolved to form the underlying basis for all life. One doesn't need to study chemistry to understand and appreciate this. We don't often consider the basic fact that much of our lives are the result of specific chemistries interacting with other chemistries.

Once the chemical elements emerged, there also emerged the first chemical equilibrium between chemical states. This could have been something as simple as an acid-base equilibrium, or a structural change between two interconnected chemical states. Then began the emergence of more complex chemical networks that interacted with these earlier chemical equilibria. What these early chemistries were, we can only guess. However, the physical forces such as gravity also emerged from the early universe and eventually formed the first stars, then planets and our sun and Earth. This greatly complicated the universe's nascent chemistries. Other phenomena also emerged during this time, leading to the formation of the heavier chemical elements within stars. That's an extremely brief synopsis of the first nine or ten billion years of our known universe.

The emergence of the elements in the universe began from the physics that provided the chemical elements necessary for all life to exist (see Appendix II). There are about 118 different elements currently in the periodic table. Several elements are only found in laboratories and nuclear accelerators. You might wonder how many elements can be found in nature. All the elements up to element 92 (uranium) can be found in nature. However, it turns out that there are also other elements naturally occurring in trace amounts that bring the current number of naturally occurring elements to about ninety-eight. This means that there are about twenty elements not previously found in nature that humans have created.

If life did develop from these chemistries on early Earth, then there is at least one set of essential chemical elements that produced the earliest life form. In whatever way life developed, we now understand that it depends on chemistries to function. For life to begin, it may have been necessary to have only 20 or so of the most commonly found elements in the earth's crust, sea water, and the human body (as listed in Appendix II). It is also interesting to note that volcanic ash contains many of these same elements.[2] The reason we're going through this exercise is to demonstrate the enormous possibilities for, and scientific challenges to, finding a suitable chemistry for life's beginning. Things that initially seem simple become much more complex the more one tries to understand them. This is how science progresses.

---

2   *Wikipedia*, s.v. "Volcanic ash," https://en.wikipedia.org/wiki/Volcanic_ash; and also Adam P. Johnson et al., "The Miller Volcanic Spark Discharge Experiment," *Science* 322, no. 5900 (October 17, 2008): 404, http://science.sciencemag.org/content/322/5900/404.

Innocently enough, we might think we know all the essential chemical elements for life, or at least those for human life.[3] The truth is that we don't. These experiments are extremely difficult or nearly impossible to do for various reasons, such as trying to control all the food and water intake of an organism and the entire living environment, ensuring that additional trace elements are completely lacking. One might think that by using cells in culture, we could determine exactly what the essential elements might be, at least for those cells growing in a defined culture medium. There are still problems with using a serum-free or chemically-defined culture medium.[4]

One of the main problems is that cells growing in culture usually require various growth-stimulating factors that might inadvertently include trace elements that turn out to be essential for life. Another confounding factor is that too much of a trace element may be toxic at larger amounts. This means that one should know the dose-response curves for these elements considered to be essential. Even the use of various buffers to maintain the acid-base balance in cell culture may contain trace impurities. Such experiments are enormously difficult to do because there are many variables that can't be adequately controlled in these studies. In addition, the arch of life's evolution has most likely altered the number and type of essential micronutrients and trace elements that are necessary for various organisms to exist. The trace elements necessary for humans may be different from those for a deep-sea fish or a bacterium. Determining the trace elements essential for life is much more complicated and less certain than most people realize.

What are common to all known life are these common elements: carbon, hydrogen, oxygen, nitrogen, sulfur, and phosphorous. Among these six elements, carbon is perhaps the most special since it can form bonds with itself and make molecules that have many different shapes. Carbon molecules can be short chains, long chains, bent chains, branching chains, and ring shapes. The four classes of macromolecules that make life possible

---

3   Prinessa Chellan and Peter J. Sadler, "The Elements of Life and Medicines," *Philosophical Transactions of the Royal Society A: Mathematical, Physical and Engineering Sciences* 373, no. 2037 (March 23, 2015): 1–56, doi:10.1098/rsta.2014.0182, https://www.ncbi.nlm.nih.gov/pmc/articles/PMC4342972/.

4   *Wikipedia*, s.v. "Chemically defined medium," https://en.wikipedia.org/wiki/Chemically_defined_medium.

(proteins, carbohydrates, lipids, and nucleic acids) are all made of carbon. Aside from the big six mentioned above, the next major elements for life to exist are sodium, chlorine, potassium, calcium, and magnesium. Considering just these basic facts brings us to eleven elements, that are present in all known life: 1) carbon, 2) hydrogen, 3) oxygen, 4) nitrogen, 5) sulfur, 6) phosphorus, 7) sodium, 8) chlorine, 9) potassium, 10) calcium and 11) magnesium.

In humans, scientists believe that about 19 of the approximately 118 known elements are essential. An essential element is one whose absence results in abnormal biological function. The other probable essential elements for life are the trace elements (minor elements), which are present in very small quantities. An element is considered a trace element when its requirement per day is below 100 mg. A deficiency of these elements is rare but may prove fatal. Examples include copper, iron, zinc, chromium, cobalt, iodine, molybdenum, and selenium. With these eight elements and the eleven elements mentioned above, we arrive at the nineteen elements considered essential for human life: 1) carbon, 2) hydrogen, 3) oxygen, 4) nitrogen, 5) sulfur, 6) phosphorus, 7) sodium, 8) chlorine, 9) potassium, 10) calcium, 11) magnesium, 12) copper, 13) iron, 14 zinc, 15) chromium, 16) cobalt, 17) iodine, 18) molybdenum, and 19) selenium. There are some additional trace elements, but their role is yet unclear. Examples include cadmium, nickel, silica, tin, vanadium, and aluminum.[5] Welcome to our vast sea of chemistries!

This was the beginning of life's chemistries. Some of the interesting developments along this path diverged from the common chemistries that we observe today. One of these diverging paths produced a group of twenty amino acids that are encoded by the genetic codes that make the proteins in all life. These twenty amino acids were somehow selected to be in the L (*levo*, or left-handed) form with little or no D (*dextro*, or right-handed) form, which means that they have basic chemical structures that are unique to all life. Normally, if we made amino acids in the lab, they'd be a mixture of

---

5   Preeti Tomar Bhattacharya, Satya Ranjan Misra, and Mohsina Hussain, "Nutritional Aspects of Essential Trace Elements in Oral Health and Disease: An Extensive Review," *Scientifica* 2016 : 1–12, doi:10.1155/2016/5464373.

L and D forms, but, for some reason, life chooses the L form of amino acids. We continue to speculate why this happened, but we don't know with any certainty why it did.

Was this a form of chemical evolution? We don't usually use the word *evolution* with chemistry, but we have observed that geological and organic chemistries can change or evolve over tens of thousands or millions of years.[6] We don't know if they can naturally create pre-life forms, but there are some promising leads.

EMERGENCE OF METABOLISM

Serpentinization is a geological process involving reactions between water and the rocks of the Earth's lower crust and upper mantle that may have been important reactions for the earliest life.[7] Serpentinization involves the elements iron, silicon, magnesium, hydrogen, carbon, and oxygen. In addition, various sulfide chemical compounds, utilizing the very versatile element sulfur, may have been vitally important for life's beginnings, especially in a chemical pathway that utilizes methane (natural gas) as a primary energy source.[8] Along with other important elements, which include nitrogen, phosphorous, calcium, magnesium, sodium, chlorine, and potassium, the chemical dance toward life began roughly four billion years ago.

---

6   *Wikipedia*, s.v. "Mineral evolution," https://en.wikipedia.org/wiki/Mineral_evolution.

7   M. J. Russell, A. J. Hall, and W. Martin, "Serpentinization as a Source of Energy at the Origin of Life," *Geobiology* 8, no. 5 (2010): 355–371. Various sulfide chemical compounds may have been important for life's beginnings. See G. Wächtershäuser, "Evolution of the First Metabolic Cycles," *Proceedings of the National Academy of Sciences (PNAS)* 87 (1990): 200–204, https://www.ncbi.nlm.nih.gov/pmc/articles/PMC53229/; for a discussion of the ancestral methanotrophic pathway, see Michael J. Russell et al., "The Drive to Life on Wet and Icy Worlds," *Astrobiology* 14, no. 4 (April 1, 2014): 308–343, doi:10.1089/ast.2013.1110; Wolfgang Nitschke and Michael J. Russell, "Beating the Acetyl Coenzyme A-Pathway to the Origin of Life," *Philosophical Transactions of the Royal Society B: Biological Sciences*, 368, no. 1622 (July 19, 2013) https://doi.org/10.1098/rstb.2012.0258. (Michael's favorite).

8   Sushil K. Atreya, "The Mystery of Methane on Mars and Titan," *Scientific American* January 15, 2009, https://www.scientificamerican.com/article/methane-on-mars-titan/.

Protocells

Protocells (also called micelles, among many other terms used to describe similar entities) are membrane-enclosed combinations of various chemical compositions.[9] These pre-life forms have the potential to evolve into life and are thought to have existed before the formation of a living Self. They must have had a long period of evolution before they became the first cell (Self). The first life could have evolved from any one of a myriad number of possible chemical combinations, which is why it is almost impossible to identify with any certainty the combination that led to the first life. Some of the experiments to identify a possible chemical origin to life have produced very interesting phenomena that appear to be at least on the road to life. It is difficult to tell if life has formed in these proto-cellular, chemical systems because we don't have a specific definition for life, which is something this book may at least begin to correct. Even in failing, we might take a step or two toward a better understanding of life in general.

In the beginning of our planet, for a period of about a half billion years, Earth's early chemistries eventually began to form membranes that could contain the internalization of external chemistries. The formation of these enclosed membranes also allowed for the development of new chemistries that could be energized by potential energy gradients across the new membranes. In addition, these membranes shielded the internalized chemistries from sudden changes in environmental conditions, such as pH and osmotic changes that would normally disrupt these chemistries in the external environment. Given these developments, the early protocells appeared to be on the road to a primitive, chemical evolution, metabolism and homeostasis.

If we begin with the origin of the most primitive Self arising from a set of chemistries, we see that some sort of membrane, serving as a barrier, was needed to separate the Self from its environment. This membrane may have been much different from the membranes in the cells we know today. It may have been largely proteins with relatively few lipids, resembling today's virus membranes. This primitive Self may have had a primitive type of memory

---

9   Wikipedia, s.v. "Protocell," https://en.wikipedia.org/wiki/Protocell. And also Irene A. Chen and Peter Walde, "From Self-Assembled Vesicles to Proto-cells," Cold Spring Harbor Perspectives in Biology 2, no. 7 (July 2010): a002170. doi:10.1101/cshperspect.a002170.

based largely on the copying of template molecules (recursive structures serving as a primitive form of heredity). Inside it must have contained primitive enzymes that could repair the membrane and ensure that it could replicate itself. It was also most likely semipermeable, which would create the Essential Tensions due to the embedded chemical and physical disequilibria. The presence of these types of semipermeable membranes with the Essential Tensions would set the stage for the beginning of life.

Hydrophobic molecules would be attracted to these membranes, which would separate hydrophilic and hydrophobic molecules, leading to other Essential Tensions within this primitive proto-Self. Within these physicochemical gradients, proteins and enzymes must have driven novel chemistries leading to the problems of Stability versus Novelty. The coupling of these physicochemical gradients to external energy sources required the manipulation of these energy streams leading to stable systems that could metabolize and grow. Eventually, these primitive proto-Selves would evolve to couple Energy Flows to other molecules, containing phosphates that could be useful to store energy and drive reactions when the energy supply was low or absent.

Encountering diverse environments led to Fateful Encounters that could generate the evolution of more complex systems. This minimalist set for the physicochemical properties of the first proto-Self was most likely far different from our present-day life. There may have been no DNA or RNA, perhaps only membranes, proteins, and primitive enzymes that may have functioned to repair and reproduce membranes and direct the Energy Flow toward new proteins and enzyme functions that could enhance any of the five characteristics of life and evolution (see Appendix I B).

These earliest Essential Tensions were also vitally important in establishing and directing Energy Flows as well as life's sensory molecules. The establishment of Energy Flow was critically important in forming a metabolism. Metabolism and Self-regulation had to evolve together in a balance between the forces building more of the Self (Stability) and countering those processes that would destroy parts of the Self (catabolism).[10] In times of severe stress, when food

---

10 Antony R. Crofts, "Life, Information, Entropy, and Time: Vehicles for Semantic Inheritance," *Complexity* 13, no. 1 (2007):14–50. https://www.ncbi.nlm.nih.gov/pmc/articles/PMC2577055/.

supplies were limited, the Self might have catabolized its internal molecules to derive energy from them. If this catabolism couldn't have sustained the Self, the Self may have become dormant or dead.

Self-regulation

Life continues to grow if the Energy Flow needed to continue the metabolic cycle is directed toward sustaining the Self (anabolism and Stability). An essential criterion for this is the presence of a chemical feedback loop that can sense and respond to changes in the Self's Growth and Energy Flow. This creates an adaptive network through these feedback loops.

Once life harnessed the energy from the Essential Tensions, such as the electron transfer across pH gradients through a cell's membrane that could sustain a metabolic cycle, there came the interplay between Stability and Novelty. This involved other chemistries that were added to the original, which thereby extended it to become a set of linked chemistries. These chemistries, which became a chain of linked chemical equilibria, must have demonstrated the earliest form of feedback. This feedback also allowed for the development of adaptive responses, which began what we call Self-regulation. These recursive processes also gave rise to life's metabolism and inherent complexity.

It wasn't enough that life developed these membrane-enclosed, internal chemistries that became life's first metabolism. There had to have been a period when the earliest life had to balance many competing processes. Not doing so would mean the destruction of the Self. This period where Stability must have competed with Novelty isn't well known or studied, but we can imagine some of these earliest struggles. Life must have tried different scenarios for its survival and reproduction and failed many times along the way.

One example of that happens even today: a daughter cell's budding off, which sometimes occurs in yeast cells. Budding is a form of reproduction that doesn't require full cell division and uses only a relatively small part of the mother cell's cytoplasm and membrane. After billions of years of evolution, budding has not become the major way living organisms reproduce. This may be because budding doesn't supply the new cell with an optimized amount of secondary material inherited from the cytoplasm of the mother cell. On the individual cell level, reproduction usually occurs as a division of the mother cell to form two daughter cells. This makes several important processes possible. First, it maintains the cell's metabolic machinery. Second, it solves the problems dealing with too much continuous Growth, leading to the formation of one gigantic cell. Third, it maintains an orderly transition of inheritable material from the mother to the daughter cells. Fourth, it resets the surface area to volume ratio, which ensures better diffusion for nutrients and essential molecules to reach the internal metabolism of the cell (see Appendix V).

Self-benefit and Energy Flow

In order to understand Self-benefit, we need to hypothesize that the initial Self is more than a feasible set of autocatalytic chemical reactions. This initial Self is itself an emergent property that led to the permissive chemical environments able to develop more complex biological chemistry.

This section answers the question of what Self-benefit is. Our conclusion is that the root signal for Self-benefit is the very existence of organic chemical reactions that proceed from reactants to products. This is due to the availability of continual Energy Flow, either directly to the chemical reactants, or through catalytic agents such as enzymes in proximity to potential reactants. The signal for Self-benefit is the reaction itself and the complement, which is the cessation of existing chemical reactions, is the signal for Self-detriment, degradation, and eventually, dormancy or death.

Chemical reactants can either proceed in reaction or not, depending on the potential energy required among the reactants to reach above some threshold. The proximity of a relevant catalyst plays a huge role in determining whether a reaction proceeds or not, by changing this threshold. Although our hypothesis for the ultimate signal of Self-Benefit appears very simple on the surface, it has a huge impact on understanding the drivers of Growth in life. First, this Self-benefit signal links inanimate organic chemistry to life's original metabolic chemistry. Second, chemical circuits can form both positive and negative feedback loops that contain both a metabolism and Self-regulation. Third, the energetics of these essential metabolic chemical reactions are fundamental to all life, which include patterns of Energy Flow at all levels. Those chemical reactions that interpret the chemical signals from these Energy Flows were the origin of determining Self-benefit in life.

Jumping ahead to life processes, to survive, life needs to find nourishment and avoid adversity. All life has sensory molecules, called receptors, that detect either harmful or helpful environmental stimuli. How did living organisms distinguish between harmful or helpful stimuli billions of years before any neural circuits existed? How does life inherently determine what is Self-beneficial in the environment and what is Self-detrimental? The most primitive motile behavior in which life determines Self-benefit is chemotaxis, by which bacteria orient toward positive gradients of nutrients. However, the known mechanisms of chemotaxis are too highly evolved to be considered the origin of Self-benefit. The origin needs to lie in inanimate organic chemistry and have continuity with the processes in life today.

EMERGENCE OF SELF

The Self is that primary, atomistic core concept of life and evolution, which should be as true for a bacterium as for a human being. We argue that a Self includes all relevant biological cells but may not be a single cell under some interpretations. We humans have emerged from the microbial world as a complex network of cells with networks connecting them to many other cells as well as the microbiological prokaryote cells that are contained within our bodies.[11]

As mentioned in chapter 1, the 1994 "Workshop on Self-Determination in Developing and Evolving Systems," the most important consensus to emerge at the workshop was "that the study of the emergence of *Self* across life forms has enormous potential for understanding phenomena in the life sciences and solving problems in computation." What are we to do with this concept? Any Self is connected in many ways to its environment; some Selves may almost dissolve into their myriad of environmental connections. The Self may also undergo dramatic and unforeseen changes.

The Self is one essential concept for our definition of life, and it contains the other four concepts of life (see Appendix I B). If the Self is unique and fails

---

11 Ed Yong, *I Contain Multitudes: The Microbes within Us and a Grander View of life* (New York: Harper Collins, 2016).

to reproduce, it may be an evolutionarily dead-end. This allows for the fact that there may be life that is unique and doesn't reproduce but can still be alive by our definition. The Self may experience varying degrees of evolutionary success based upon many complex factors, including its interactions with others of the same or different kinds and new environments. The Self may cease to exist if it can no longer integrate or maintain one or more of the other four essential concepts that we've discussed.

The Self also displays behaviors based upon its reactions to the absolute amounts and types of stimuli. This provides it with feedback about the environment. In a real sense, life is the coupling of positive feedback and negative feedback loops into a network that controls Energy Flows. These intermediate negative feedback loops are comprised of enzymes that couple with additional factors that regulate enzymes and proteins within a cell. At their core, they function largely as multistate systems to create complex, nonlinear activities.

The Self and Composite Organisms

There are many composite organisms composed of many other life forms. We've discovered that our own human cells are only a fraction of the total cells in our bodies! We also have a metabiome that links to our brain-gut connection and other microbial life. There are many other creatures besides us that are composite organisms. Most life on Earth functions together with a multitude of other organisms that we're just beginning to understand and explore.

The Self requires the internalization of external chemistry. As a result, it also developes a complementarity between the Stability of Self and the Novelty of possible chemistries driven by the energy gradients across the membrane. The membrane also selectively allows some molecules internal access (Fateful Encounters) that add to the Self (amino acids, proteins, RNA, DNA, etc.). For this Self to continue to develop, it must divide to retain the dynamic physical-chemical tensions that drive the energy gradients by making new membranes. This would have been a Self-benefit that allows replication. Initially, this may or may not have occurred with an informational molecule such as RNA. This is the adaptive regulation of Stability versus Novelty revolving around the Stability of this new membrane-encapsulated Self and the Novelty of new internal chemistry driven by membrane energy gradients. These are the general concepts applied to the chemistry of earliest life.

Because protocells have a clear boundary from their environment with their membranes, the protocell becomes the first entity we can say has a primitive pre-Self, but is not a living Self. Concomitant with the Self, there also must be the *other* which would be its environment and other Selves. So how do the dualistic processes on creative tensions inherent in the Self lead to gaining complexity and, ultimately, to the properties of life? To answer this question, we first need to clarify whether there are any unifying drivers for all these processes.

EMERGENCE OF LIFE

According to some scientists, a useful operational definition of life is "a self-sustaining chemical system capable of Darwinian evolution." This definition was adopted by the exobiology program at NASA.[12] This begs the questions of what a self-sustaining chemical system is and how we would see evolution within this system.

The emergence of life and the Self is only recognized in the context that life had solved the initial dilemmas of continuous Growth and the necessary balances between Stability and Novelty. These required the development of inheritance of the metabolism necessary for maintaining the Essential Tensions, which were incorporated into the permanent structures of the Self. The existence of a pre-Self with a primitive metabolism that includes a mechanism to maintain or synthesize the membrane begins the physicochemical road to life. We might pause here and ask what is lacking. One important concept that is lacking is the ability to grow and reproduce. Any ability to grow larger must also entail an ability to reproduce. Otherwise we wind up with just one giant cell. If life had started down this path, it must have discovered that one giant cell wasn't the most stable system.

Using our five concepts, we attain a more detailed definition of life

---

12 Irene A. Chen and Peter Walde, "From Self-Assembled Vesicles to Proto-cells," *Cold Spring Harbor Perspectives in Biology* 2, no. 7 (2010): a002170. https://www.ncbi.nlm.nih.gov/pmc/articles/PMC2890201/.

involving the progression of protocellular forms into more life-like forms. Protocells are in a state of flux in terms of most of our five essential concepts of life. The concepts that may be most in a state of flux are the Essential Tensions versus non-Essential Tensions, Growth and Energy Flow versus stasis (dormancy) and no Energy Flow, and Stability versus Novelty. These concepts involve the selection of a proper balance between competing chemical and physical processes and forces. Eventually, this development should lead to the reproduction of a living Self by having a minimal metabolic set that can copy the internal proteins and enzymes necessary for maintaining metabolism and the cell membrane.

Death emerges as the removal of Essential Tensions. As these are removed, the Self loses its Energy Flows. This imposes normal equilibrium conditions on those biochemical processes that were embedded as structural and chemical disequilibria. Death returns us to the basic chemical world; we've emerged from the chemistries of the universe and they take us back...

THE EMERGENCE OF INFORMATION

Information is everywhere. Information is embedded in the universe, but we recognize it only when it is interpreted within a larger context. The term *information* may mean different things under different contexts. Is information only found within the existing literature on Earth? Many might answer no, but then, what is it? Is it an ordering of matter? If so, what constitutes an ordering of lower entropy? Is it only when we can discern a pattern? Many believe that the molecules of life, DNA and RNA, harbor information that describes an organism. Is this true, or are other molecules necessary to give order to and expression to the DNA and RNA molecules? Information by its very nature is embedded within the context of its environment. There is no such thing as information alone.

Information may be an important construct of our minds that assists us in compiling facts that lead our scientific investigations forward. Information resides within our five essential concepts as the structural constraints imposed upon each of the essential criteria in the list. The constellation of these five concepts constrain biological patterns upon their underlying chemistries.

We organize rather extensive groups of observations into categories that assist us to understand the underlying reasons behind scientific facts and how
they relate to other facts. This is the basis for scientific inquiry. Depending how
close we shave with Occam's razor, our scientific calculations, observations,
and theories can occupy large portions of our knowledge. It is interesting to
speculate how much larger this knowledge base would be if we still had the
older Aristotelian and Ptolemaic systems as the basis of our current scientific
knowledge. Because these systems have been refined and surpassed by simpler
and more accurate scientific descriptions, our total information load is thereby
relatively smaller. This suggests that our current information is relative to the
accuracy of our current scientific theories.

Somewhere there may exist such a thing as an environmental battery, which
represents a separation of charges by a barrier that prevents spontaneous equilibrium between sets of separated charges. The concept of a battery and how
to design one is one type of information. We might postulate that because the
separation of charges arose from the environment *de novo*, information has
emerged from this environment. Has the information emerged, or have we just
become aware that it is possible that types of batteries may form within the
numerous possibilities of potential combinations of environmental elements?

Some have suggested that thinking about information can transform the
way we view life.[13] Instead of trying to recreate the chemical building blocks
that gave rise to life billions of years ago, some evolutionary scientists look at
key differences in the way that living creatures store and process information.[14]
One part of this problem is that information often corresponds with our concepts of order, which appears to contradict the second law of thermodynamics
that the total entropy of an isolated system can never decrease over time.[15]

---

13 Antony R. Crofts, "Life, Information, Entropy, and Time: Vehicles for Semantic Inheritance," *Complexity*
    13, no. 1 (2007): 14–50, https://www.ncbi.nlm.nih.gov/pmc/articles/PMC2577055/.

14 Tia Ghose, "Origin of Life Needs a Rethink, Scientists Argue," Live Science, December 11, 2012,
    https://www.livescience.com/25453-life-origin-reframed.html. See also Christoph Adami, "What
    Is Information?" Philosophical Transactions of the Royal Society A: Mathematical, Physical and
    Engineering Sciences, http://doi.org/10.1098/rsta.2015.0230; Christoph Adami, Information-Theoretic
    Considerations Concerning the Origin of Life," *Origins of Life and Evolution of Biospheres* 45, no. 3
    (September 2015): 309–317, https://doi.org/10.1007/s11084-015-9439-0.

15 *Wikipedia*, s.v. "Second law of thermodynamics," https://en.wikipedia.org/wiki/Second_law_of_
    thermodynamics.

Entropy

Our intuitive sense that living beings must have a low state of entropy may not square with the thermodynamics of life. Overall, the net energy expenditures to create the first life on Earth must have been a net positive entropic event. Even some of the simplest, yet lower, entropic forms, such as mineral crystals, had to form under net positive entropic conditions. One reason we frequently stray from this view is that we think of life and our civilizations as lower entropic forms when in fact, if we calculated all the energy sources put into them, we'd discover that a great deal of heat was generated in their formation and thus overall positive entropy was produced. Our perceived order does not always represent the true entropic order or a decreased entropy. Cells, cities, and buildings appear to be ordered in our simplistic observations but are high entropy forms of matter. As an example, the building of a skyscraper requires the input of enormous amounts of energy to put the required materials together. The total amount of all the energy required increases the overall entropy even though the final product appears to have less entropy than its surroundings. So, too, with life, where we see order and structure created from large amounts of energy supplied from the environment, the overall entropy increases.

Living beings aren't entirely isolated systems unless they are in a completely dormant state, such as spores or seeds. The fact that life can become dormant suggests that life is a structurally based system. Thermodynamically there is very little or no difference between dormant and dead. We argue that life when viewed in totality creates more entropy and is therefore thermodynamically favored. Just add water to the seeds or spores to see which ones can grow!

Receptors as Information Vehicles

Because receptors are the way life acquires information from the environment, this is an overlooked area of information that plays a very important role in evolution. Life channels Energy Flows into molecules that can function as chemical modulators and mimics.[16] In the fascinating case for odorant

---

16 Sandra Steiger et al., "The Origin and Dynamic Evolution of Chemical Information Transfer," *Philosophical Transactions of the Royal Society B: Biological Sciences* 278, no. 1708, (2010): 970–979, https://www.ncbi.nlm.nih.gov/pmc/articles/PMC3049038/.

(olfactory) receptors, there are multiple types of these receptors that detect a broad range of chemical molecules. An effective odor cue is often triggered by some subgroup of these receptors interacting with at least one odor molecule. These detection systems involve subgroups of receptors being modulated to varying extents by the odor molecule. Modulation means that as these receptors turn on and relay their response to the cell's cytoplasm, these responses may vary through a large range of possible magnitudes including negative receptor responses.[17] This system appears to be very similar to the immune system that detects a foreign intruder and develops antigens to combat it and all of our sensory receptors.

Net Shift

One of the most basic kernels of information is the net shift in a chemical equilibrium that underlies all our receptors and sensory perceptions (see Glossary and Appendix III A, B, and C).[18] These net shifts are what drive our senses, enabling us to perceive our world and, subsequently, adapt to changes. They underlie all our perceptions and form the basis for what we call *information*.

EMERGENCE OF HEREDITY

Organisms that fail to adapt to their environments eventually cease to exist. In order to adapt to environmental changes, organisms must have mechanisms that pass adaptive traits on to the next generation. Heredity isn't mentioned as one of our fundamental criteria for life (see Appendix I B), yet it is one of the most important for life to continue, grow, and evolve. We are taking the position that heredity is an emergent phenomenon that arose from our five

17 Richard G. Lanzara, "Method for Determining Drug-Molecular Combinations That Modulate and Enhance the Therapeutic Safety and Efficacy of Biological or Pharmaceutical Drugs," US Patent, US20090012717A1, July 5, 2007.

18 Richard G. Lanzara, "Weber's Law Modeled by the Mathematical Description of a Beam Balance," *Mathematical Biosciences* 122, no. 1 (July 1994): 89–94, http://www.bio-balance.com/Weber's_Law.pdf. And also *Wikipedia*, s.v. "The Weber–Fechner Law," https://en.wikipedia.org/wiki/Weber%E2%80%93Fechner_law.

essential concepts for life. This is a controversial position, but just as mito-chondria, chloroplasts, and nuclei have emerged as structural elements, so, too, have our genetic biochemistries emerged. All our current genetic heredity is, essentially, the underlying chemistries of genes, DNA, RNA, and the accompanying proteins and molecules that support and control them. These chemistries arose from the adaptive chemistries of the past.

The Self may have at least one set of memory molecules (DNA, RNA, etc.). Although there is a wealth of evidence to support the importance of DNA and RNA, a problem arises when we consider the earliest life forms: it is difficult to imagine that the supporting proteins and molecules required to create DNA and RNA were present in the earliest life. This is much like trying to imagine how a jet would fly if we know nothing about planes, or even if we do know something about planes: how would a jet, weighing tons, even get into the air? Were there alternative molecules that functioned as repositories of inheritance?

The Self has multiple ways to convey hereditary information about its past state to new Selves. Certain structural constraints, such as those constraints found in crystals and minerals, provide template memories that don't require specific memory molecules to convey structural organization. Membranes are natural sensory/memory repositories. Production of new membrane upon normal cell division is an overlooked place for the features of Novelty, evolution, learning, and environmental feedback (since each daughter cell inherits about 79 percent of its membrane as old membrane from the mother cell and 21 percent as newly formed membrane (see Appendix V)). This may become a "good" (or benefit, which may include Michael's concept of Self-benefit) for the Self if a part of the new membrane formation for dividing cells obtains memory-preserving structures from the older Self's membrane and protein structures.

Perhaps the most prominent criteria for heredity are the criteria of Growth and Energy Flow, and Stability versus Novelty. We maintain that as the criteria for Growth and Energy Flow, and Stability versus Novelty became larger, more complex, and difficult to store as relatively simple templates, those molecules that could assume roles as information molecules became ripe for emergence as the repositories of protein information. This suggests that a progression of

changes occurred to the earliest life that probably didn't have the full complement of hereditary and information molecules that we see today. Earliest life
must have had the most rudimentary tool box for storing hereditary information. These earliest hereditary systems may have relied on the templates of the
already-existing proteins that functioned to store hereditary information. This
may have also included catalytic forms of RNA molecules (or ribosomal RNA)
such as Ribozymes (ribonucleic acid enzymes) that are capable of catalyzing
specific biochemical reactions, similar to the action of protein enzymes.[19]

Heredity may come from the genetic material of an organism, yet it also
has a significant component in the epigenetic and/or structural components
within the cells of an organism. Both mitochondria and chloroplasts contain
their own DNA. In microorganisms, there are also multiple ways that genetic
transfer can occur. These include transfer by viral means as well as direct transfer from one microorganism to another. It is just beginning to be recognized
that these alternative ways to transfer genetic material may also be operating
in the higher life forms such as the eukaryotes.

Looking only at molecules embedded in the earliest cell's membrane,
these were molecules of protein made from amino acids. The amino acids
have many interesting and varied properties that when combined to form
a protein, give that protein, in turn, interesting properties. Some proteins
don't favor being in a watery environment. They prefer the lipids found in
cell membranes. Parts of other proteins might favor the membrane environment, or the external environment, or the inside environment of the cell.
Some parts of the larger proteins may exist in all three environments. Building
upon these basic properties of proteins suggests that the early protocells that
could reproduce successful sensory molecules could also successfully evolve.
This represents a possible chemical pathway to life, which must have initially
relied on some type of primitive inheritance based upon the template structures of these proteins and membranes. Although this form of heredity may
not fit with our current understanding, earliest life had perhaps more than
a billion years to evolve the more sophisticated forms of heredity that we
observe today.

---

19 *Wikipedia*, s.v. "Ribozymes," https://en.wikipedia.org/wiki/Ribozyme.

Feedback

We also want to know how life receives feedback from change and how it stores that feedback as inheritable material. You've received not only mitochondrial heredity from your mother's egg cell but also the cytoplasmic, epigenetic, and membrane heredity from your own original egg cell. Environments can change how our development occurs. Sometimes this is dramatically illustrated by how some chemicals can cause birth defects.

The First Gene

The necessity for top and middle management becomes necessary for most growing entities (biological, socialogical, and corporate). There came a time in evolution when life needed to include some form of management. This is also evident in the social hierarchy of many animals. So where does this come from? It probably begins with some form of mimicry or peripheral involve-ment. If you are subtly influenced by some activity or other, you may find your-self drawn into the action and begin providing instructions, like a coach who doesn't compete but instructs others on competing. Suddenly, a new career is born that didn't exist before. So, too, on the chemical level, there can develop molecular catalysts or molecular modulators that alter the speed, direction, and amount of reaction. On the level of life's chemistry, life had to have had some influence on the chemical witnesses to this process. These molecular witnesses may have originally been proteins or forms of catalytic or ribosomal RNA[20] within the cell that were caught up in complex feedback loops..

Earliest life was also developing transmembrane proteins that could protrude from the external membrane. The formation of such proteins was complicated by the fact that proteins are often difficult to insert through a membrane. Building these transmembrane molecules involved every part of the cell, including the cytoplasm and membrane. This may be analogous to

---

20 R. Root-Bernstein and M. Root-Bernstein, "The Ribosome as a Missing Link in Prebiotic Evolution III: Over-Representation of tRNA- and rRNA-Like Sequences and Plieofunctionality of Ribosome-Related Molecules Argues for the Evolution of Primitive Genomes from Ribosomal RNA Modules." *International Journal of Molecular Sciences* 20, no. 1: 140, https://doi.org/10.3390/ijms20010140, https://www.mdpi.com/1422-0067/20/1/140.

building a rocket ship to go to the moon. Originally, rocketry was a hobby pursued by one or two individuals, but when it grew, it required teams of individuals to design and make a rocket. So too with early life, the first genes must have been caught in this manufacturing feedback loop between the metabolic pathways and the transmembrane proteins of the cell.

Genes as Feedback Mechanisms

Genes are elegantly poised to do their timely dance at the beginning of an organism's life. They turn on and off in an intricately timed manner that puts any Swiss watch to shame. This process is called embryogenesis and it is how every organism that grows from an egg cell or seed is formed. It is the dance of the genes.

We know that only about 1 percent of human genes have no similarity with the genes of other animals. This means that our genetic similarity to other animals is enormous and demonstrates that a relatively small number of genetic alterations can create large changes in an organism's phenotype, which determines how an organism functions and appears.[21]

Genes are vitally important to all life and are enormously complicated. As an analogy, consider that you're playing poker with each card representing a gene and each hand representing an organism. Now let's say you start playing, but you're not sure of the rules. Are you playing Texas hold'em or five-card draw, or what? Now we find out that the deck has two jokers and the jokers allow you to double or even triple the number of cards in your hand. Then we find out that twos and fives are wild. Next, we find out that we're playing two poker games at the same time! These extremely complex biochemical entities are only just beginning to be understood. In many ways, genes are the biochemical conductors that tell our cells when and how to manufacture proteins that then modify the activities of other proteins as well as other

---

21 Manyaun Long et al., "The Origin of New Genes: Glimpses from the Young and Old," *Nature Genetics* 4 (November 2003): 865–875, http://www.bath.ac.uk/bio-sci/hejmadi/BB30055/gene%20evol%20 nrg1204.pdf. See also Margarida Cardoso Moreira and Manyuan Long, "The Origin and Evolution of New Genes," *Methods in Molecular Biology* 856 (January 2012): 161–186, doi:10.1007/978-1-61779-585-5_7, https://www.researchgate.net/publication/221686019_The_Origin_and_evolution_of_New_Genes.

genes. This is a vitally important type of feedback that we're just beginning to understand.

There are many ways in which genes can be altered to become silent or to create a new gene. Genes appear to exist in large and complex biochemical networks that create conditions for change and, ultimately feedback to generate change within the genes. The grand interplay of these complex networks creates conditions for life's evolution. What may be most surprising is that life may change, combine, and alter genes between species. Genes are not the static entities that the early biologists thought they were. A gene's transformation into a different gene may not destroy the previous function of the old gene. The new gene may also provide more protein diversity and may also affect the expression of other genes. Both genes and proteins probably have more than a single function, which can greatly complicate the scientific analysis of these events. This is a relatively new area of study. We have much more to learn. What we're witnessing is just the beginning of fully understanding what we call genetic and epigenetic inheritance.[22]

EMERGENCE OF EVOLUTION

We're all connected to the first life and our chemistries are related to all life's chemistries existing on our planet today. The first life was, in its most basic form, a cytoplasm enclosed by a semipermeable membrane. A necessary and emergent property of this early life was the chemistry that could repair and replenish the membrane if it became damaged. Once these basic chemistries were established, the most amazing emergent property that appeared in the evolution of life was perhaps life's ability to move. This allowed the newly emerged Self to search for better environmental conditions.

Let's pause for a moment and contemplate what this means. If we anthropomorphize, we might say that, somehow, this newly formed Self recognized that it wasn't in its optimal environment. It was unhappy! It knew it could find something better. So, it began to move away from its current subpar

---

22 Henrik Kaessmann, "Origins, Evolution, and Phenotypic Impact of New Genes," *Genome Research* 20, no. 10 (2010): 1313–1326, doi:10.1101/gr.101386.109.

environment toward one that might be better. On the biochemical level, many events were taking place. First, this new Self had to have a way to sense that it wasn't in an optimal environment. This may have been something as simple as a change in the pH-gradient (hydrogen ion concentration or acidity). A change in the environmental pH might mediate the effect of one or more internal metabolic pathways. In order to produce some means of movement, this could have been linked to an increased fluidity of the membrane causing the formation of a primitive pseudopod, similar to how amoebas move today. That early Self's movement away from a harmful (or toward a more beneficial environment) was an incredible emergent phenomenon that we don't often recognize. Thus began the evolutionary journey of life up to our present day—truly a mind-blowing thought!

Change must precede evolution but is also an essential part of it. Yet in nature there are examples of organisms that have not changed for many thousands or millions of years—for example, sponges, coelacanths, crocodiles, and horseshoe crabs, to name a few. Evolution has been described as a process of change, which is thought to be necessary for life. What if that is wrong? What if some organisms don't evolve? Are some organisms in an evolutionary dead-end? What if the processes of change are reduced to specific biochemical concepts? We begin by listing some of the types of change that might drive life's evolution:

- *Sources and drivers of change.* Fateful Encounters and Novelty drive changes through attractive interactions that may improve survival.
- *Regulation of change.* Stress and pain drive change through feedback to maintain survival.
- *Potential for change.* Energy Flow is the basic currency of survival and change.
- *Context of change (Essential Tensions).* Change can occur within the Self, within a group, or across peers, such as cooperation versus competition.
- *Propagation of change.* Propagators of change include copying into inheritable material (both template and structural), information molecules, feedback loops in metabolism, and sensory receptors.

Based on the five essential concepts of life, we can see what processes must be the most important. First, the Self must be retained. This is one reason that chloroplasts and mitochondria aren't considered as independent living entities because they long ago merged with their host cells to form new eukaryote cells and thereby gave up much of their metabolism and heredity to their host cells. Second, the Essential Tensions necessary to maintain an independent Self must be preserved. Third, Growth and Energy Flow along with Stability versus Novelty are requirements for the Self to eventually reproduce. The Self's ability to reproduce solves several problems with continuous Growth, which would eventually lead to instability. In the absence of Growth and Energy Flow, the Self might enter dormancy. The Self may be dormant but still alive because it contains the structures necessary to engage in the other four essential concepts. Fourth, Stability versus Novelty requires the balancing of the metabolic cycles of the cell to produce useful energy or engage in cycles that are not producing useful energy but could lead to new chemical branches of a Self's metabolism (Novelty). Fifth, Fateful Encounters may occur at any time, even during dormancy, since irradiation of seeds can alter the subsequent development of plants, to name one example of how a Fateful Encounter might produce change in evolution and life.

Applying the Five Essential Concepts of Life to Evolution

1. The *Self versus the non-Self.* This refers to the concept of a Self or an organism as applied to an entire species.
2. *Essential Tensions versus non-Essential Tensions.* These are sets of imbalances that separate species from each other.
3. *Growth and Energy Flow versus stasis (dormancy) and no Energy Flow.* This refers to the Growth required to maintain a species and/or evolve a new species.
4. *Stability versus Novelty.* There is a dynamic balance between the stability-enforcing power of a species and the novel expansion of physical and chemical spaces for potentially new species.
5. *Fateful Encounters versus no Fateful Encounters.* Potential expansion results from new physical, chemical, and biological encounters in the life

of a group of organisms or species. Some encounters may cause harm or competition among different species. These may also lead to mergers (e.g., some parasites), or symbiosis, which is change by interdependence or union or eating or being eaten. These encounters depend on the type and number of environments a species occupies. They can also happen within a species, which may lead to Novelty and eventual Stability, as mentioned above. With these five concepts, we attempt to forge a more detailed description of life's beginnings and continued evolution.

Earliest Life

For the first 2.5 billion years or so of early Earth's history, the earliest life existed mainly as prokaryote and archaea cells. Prokaryotes and archaea don't have a cell nucleus, mitochondria, or chloroplasts. Due to the absence of a cell nucleus, the cell's DNA floats within the cell's cytoplasm. This is one of the least understood areas of evolution because fossil evidence is scanty. For a long period of time, only microbial life existed. Our earliest records come from fossilized stromatolites, which appear to be over three billion years old.[23] Discovering the oldest rocks in the hope of finding evidence of the earliest life is extremely difficult, yet it is still an active area of research.[24]

Which form of earliest life could have led to the more modern eukaryote cell? Archaea appear to have a more diverse range of metabolic tools that might have formed the basis for the current forms of metabolism observed in our eukaryote cells today. At first, archaea were thought to be only extremophiles, which is to say that they were only found in extreme environments such as hot volcanic springs. However, they are now known to be everywhere and have a very broad range of metabolisms.[25] These different metabolisms developed in many extreme environments, using everything from hydrogen sulfide, ammonia, methane, and sunlight as food.

23 *Wikipedia*, s.v. "Stromatolite," https://en.wikipedia.org/wiki/Stromatolite.
24 *Wikipedia*, s.v. "Planetary habitability," https://en.wikipedia.org/wiki/Planetary_habitability.
25 Wikipedia, s.v. "Archaea," https://en.wikipedia.org/wiki/Archaea.

Along the way, about two billion years ago, life took a dramatic turn and evolved into a much more complex organism: the eukaryote cell, which was somewhat like a large composite of the archaea and prokaryotes that had come before. This formation of the eukaryote cell might have involved one or more mergers with archaea and/or prokaryote cells, which might have been facilitated by a virus, but we aren't sure what really happened. The eukaryote cell has a nucleus that holds the cell's DNA, and a cytoplasm with many structures not found in either archaea or prokaryote cells. These structures include mitochondria, as well as other complex structures. These were some of the most dramatic evolutionary developments in the history of life because they eventually gave rise to humans and the other life forms of today. The internal structure of plants include everything in eukaryote cells as well as the chloroplast structure, which allows them to use light as a food source. We don't know how these events happened, but scientists have speculated that mergers between archaea and prokaryotes produced the eukaryote cells, which are the animal and plant cells of today.[26]

Meanwhile, life evolves on its merry way. We do know that it can produce an organism with new capabilities very quickly. It has been less than fifty years since plastics were introduced into Earth's environment, yet in a relatively short evolutionary time span, some bacterial life has evolved to eat plastic![14] In addition, bacteria once susceptible to penicillin are now resistant to it and other antibiotics. Life can evolve quickly if the situation permits it, just as workers without exactly the right tool for the job can still get the job done by jerry-rigging whatever tools they do have in their tool box. Life also has its tool box. All vertebrates, which include humans, have only about 20,000 genes that produce all the proteins needed for them to live.[15] With these 20,000 genes as the tool box, we see the panoply of evolution on view in all the vertebrates from the fishes to the reptiles to the mammals and we humans! We are only beginning to peer inside these complex boxes of life.

Life doesn't design organisms like an engineer despite what Robin Williams reportedly once said: "The human body was designed by a civil

---

26 Peter Ward, *Lamarck's Revenge: How Epigenetics Is Revolutionizing Our Understanding of Evolution's Past and Present* (New York: Bloomsbury, 2018).

engineer. Who else would put a waste plant next to a recreational area?" Many unusual looking creatures have somehow survived through evolution. One needs only to look at the big-bellied seahorse[27] or a flamingo to see this. The flamingo has an ungainly neck, and backward-curving bill and knees, which certainly don't constitute the optimal design for an efficient flying machine! There are fishes that fly and birds that swim. There are many strange-looking creatures that somehow survived through evolution, such as the dinosaurs, the deep ocean fishes and jellies, some with flashing lights like an alien spacecraft. We can only marvel at these unique beings! We're only beginning to find the many marvelous beings that were simultaneously evolving or going extinct in connection with modern humans. Life forms will continue to amaze us as more unusual species are discovered.

Sexual Selection versus Survival

Many protists (single celled eukaryotes) reproduce sexually, as do multicellular plants, animals, and fungi. In the eukaryotic fossil record, sexual reproduction first appeared 1.2 billion years ago in the Proterozoic Eon.[16] Sexual reproduction seems rather inefficient if one considers that, the male population doesn't directly bear young. Simple division would seem more efficient. If starfish are cut into pieces, they regrow. Why didn't evolution preserve a more efficient means to reproduce?

A second example of evolutionary emergence is the unexpected emergence of an animal arms race due to sexual selection pressures. For those organisms that use sex as the primary means of reproduction, there are many problems associated with this very complex approach, as any teenager will gladly inform you. Tongue in cheek aside, the fact that males must somehow compete violently with the other males to attract the attention of females, despite the dangers of doing so, doesn't seem to be an effective strategy for life and evolution. Yet male animals continue their potentially Self-destructive arms race to attract females.

_____

27 *Wikipedia*, s.v. "Big-belly seahorse," https://en.wikipedia.org/wiki/Big-belly_seahorse.

Evolutionary biologists have wondered why some animals develop massive horns or antlers when the growth of these extensions threatens the males' survival. Survival of the fittest is not the main idea here. Rather, survival of the one that can mate and reproduce is what is most important. This is an amazing emergent phenomenon! It's recognizable in a number of biological instances from the gaudy plumage of birds to the ruthless competition between males for females. The long neck of the giraffe may have developed due to sexual selection pressure since males with longer necks can more ruthlessly attack their male rivals, thereby winning the fertile female giraffe (Male giraffes attack each other by swinging their heads to strike their rivals in the side). Are we humans competing and striving for power, money, or fame today because we have this deeper evolutionary imperative that emerged eons ago as an unexpected emergent phenomenon in evolution?

What happens if the number of potential mates is reduced? Other related species may seem attractive! Such reproductive stress may account for the hybridization that occurs among different species. Hybridization also causes an increase in horizontal gene transfer (the transfer of large segments of genetic material, sometimes referred to as genomic islands).[28] During stressful conditions that threaten to kill an organism, its environment may contain many biological molecules caused by die-off. Some of these molecules are DNAs and RNAs that an organism may incorporate to effectively alter its internal milieu, including its genetic code. Under extremely stressful conditions, an organism must perform extraordinary feats of re-engineering its metabolism and genetic code. These sudden changes may be the reason some evolutionary biologists have coined the term *punctuated equilibrium* to account for the creation of new species.

---

28 Wikipedia, s.v. "Horizontal gene transfer (HGT)," https://en.wikipedia.org/wiki/Horizontal_gene_ transfer. See also an excellent article on the evolution of replication by Carl. R. Woese, "On the Evolution of Cells," *Proceedings of the National Academy of Sciences (PNAS)* 99, no. 13 (June 19, 2002): 8742–8747, http://www.ncbi.nlm.nih.gov/pmc/articles/PMC124369/. It argues that horizontal gene transfer (HGT) came before modern DNA replication and cell division. It also argues that evolution using HGT created the evolution within colonies of the first protocells. (Michael's comment: "The colony is the unit of evolution, not the individual cell." This is an idea we also talked about years ago.)

EMERGENCE OF SPECIES

From the time of Aristotle until the eighteenth century, species were seen as fixed in the hierarchy of the great chain of being. In his *The Origin of Species* of 1859, Charles Darwin explained how species could arise by natural selection and evolve over long periods of time. A species is the basic unit of classification in biology, but it is difficult to find a single satisfactory definition. A species has a set of compatible genetic elements that can accommodate genetic inheritance between members of the species. Many scientists depend upon a specific definition of species in order to understand evolutionary processes, but these definitions are fuzzy. A species is often defined as the largest group of organisms in which two individuals can produce fertile offspring, usually by sexual reproduction. This definition is often inadequate when looked at more closely, as for example, with hybridization, in which one species can successfully mate with another. This may have occurred in man's evolutionary history with the related Neanderthals and Denisovans, since some modern humans have DNA remnants from these now-extinct species.[29]

Our initial problem is that Darwin requires variation for evolution to happen. Mendel's rules only allow variation within the limits of the species gene pool, so no new species can evolve.[30] The modern synthesis points to random point mutations of bases as the driver for evolution and the selection pressure of the environment as the direction. The problem is that point mutations almost always produce either neutral or detrimental changes to the phenotype. With three billion bases in the genome and mutations at ten per generation, any beneficial mutations could not happen fast enough to explain the cluster of gene mutations in the brain, hand, and larynx of the human genome. Eighteen mutations, all clustered, invalidates the random mutations concept. What is driving the evolution of species?

The concept of a species is fuzzy because one species often interacts with another species to form hybrid organisms. Sometimes one species may

---

29 Ann Gibbons, "Revolution in Human Evolution," *Science* 349, no. 6246 (July 24, 2015): 362–366, doi:10.1126/science.349.6246.362. See also Google Scholar, s.v. "Neanderthal and Denisovan," https://scholar.google.com/scholar?q=neanderthal+and+denisovan&hl=en&as_sdt=0&as_vis=1&oi=scholart; *Wikipedia*, s.v. "Denisovan," https://en.wikipedia.org/wiki/Denisovan.

30 *Wikipedia*, s.v. "Gregor Mendel," https://en.wikipedia.org/wiki/Gregor_Mendel.

subsume another species that then appears to go extinct. This has probably occurred often in life's long history, but hasn't been well documented, because it may happen over a time span of thousands or millions of years and leave scant traces in the fossil record.

Some of the phenomena causing speciation were attributed to genetic variability from mutations and recombination, horizontal gene transfer, geographical isolation, hybridization and polyploidy (multiple copies of genes), and genetic drift with varying selection pressures. Whatever the direct causes are, the new species has a new set of inheritable expressions that set it apart from the older species from which it came. Every new species must have come from an older one. Some species become extinct, but while they exist as a species, they provide a bridge to create another new species or even multiple new species.

THE EMERGENCE OF COMPLEXITY

Complexity may arise from our inabilities to simplify the basic scientific theories that explain our experimental observations. If there are multiple layers of complexity, our minds may not be able to simplify the underlying principles. Sometimes we must wait for a breakthrough in another area of mathematics or science that relates to the areas we are studying.

As Michael mentions in chapter 4, "Quintessence," the term *selective pressure* is just a bad choice of words and meaning. In physics, pressure comes from a force. There is no force in selection. When the term *selective pressure* is used in relation to evolution, the real meaning of the term is "hardship in surviving." Natural selection does not create complexity. Complexity emerges from life's interactions of Essential Tensions with Novelty and Fateful Encounters that are iterated over thousands and millions of years.

Complexity and Complex Adaptive Systems

Complex systems can arise from simple systems that are repeated over and over (the technical term for this is *iteration*). By this process of repetition, these systems may be subject to small changes that become much greater over successive iterations. This has been demonstrated by Benoit Mandelbrot[31] and Stephen Wolfram[32] among many others who've studied chaos theory.[33] This is how complexity emerges. If we had better measures for complexity (or simplicity), then we might make more scientific progress in this area.

Wolfram's new kind of science is basically the study of simple abstract rules—essentially, elementary computer programs. In almost any class of a computational system, one very quickly finds instances of great complexity among its simplest cases (after a time series of multiple iterative loops, applying the same simple set of rules to itself similar to a self-reinforcing cycle using a set of rules). This seems to be true regardless of the components of the system and the details of its setup. In his book, *A New Kind of Science*, Wolfram explores systems that include cellular automata, Turing machines, primitive recursive functions, nested recursive functions, combinators, and several varieties of substitution and network systems. For a program to qualify as simple, there are several requirements:[36]

1. Its operation can be completely explained by a simple graphical illustration.
2. It can be completely explained in a few sentences of human language.
3. It can be implemented in a computer language using just a few lines of code.
4. The number of its possible variations is small enough that all of them can be computed.

Generally, simple programs tend to have a very simple abstract framework. Simple cellular automata, Turing machines, and combinators are examples of such frameworks, while more complex cellular automata do not necessarily qualify as simple programs. It is also possible to invent new frameworks, particularly to capture the operation of natural systems. The remarkable feature of simple programs is that a significant percentage of them can produce great

31 *Wikipedia*, s.v. "Benoit Mandelbrot," https://en.wikipedia.org/wiki/Benoit_Mandelbrot.

32 *Wikipedia*, s.v. "*A New Kind of Science*," https://en.wikipedia.org/wiki/A_New_Kind_of_Science.

33 *Wikipedia*, s.v. "Chaos theory," https://en.wikipedia.org/wiki/Chaos_theory.

complexity. Simply enumerating all possible variations of almost any class of programs quickly leads one to examples that do unexpected and interesting things. This leads to the question that if the program is so simple, where does the complexity come from? In a sense, there is not enough room in the program's definition to directly encode all the things the program can do. Therefore, simple programs are minimal examples of emergence.

Simple programs are capable of a remarkable range of behavior. Some have been proven to be universal computers. Others exhibit properties familiar from traditional science, such as thermodynamic behavior, continuum behavior, conserved quantities, percolation, and sensitive dependence on initial conditions. They have been used as models of traffic, material fracture, crystal growth, biological growth, and various sociological, geological, and ecological phenomena. Wolfram argues that another feature of simple programs is that making them more complicated seems to have little effect on their overall complexity. This is evidence, Wolfram maintains, that simple programs are enough to capture the essence of almost any complex system.

For complex *adaptive* systems, we need to add nonlinear dynamics to the mix. The term *nonlinear dynamics* describes many complex biological phenomena.[34] This particularly applies to biological receptors (see Appendix III A, B and C). In brief, it appears that almost any two-state system can function as a receptor. The active state can trigger some essential reaction that produces a larger, internal response. This usually occurs across a semipermeable membrane. The active state is triggered by a stimulus that has unequal effects on the chemical equilibrium between the active and inactive states. These stimuli can be either physical or chemical stimuli that are commonly found in most environments. They include acidity (pH), light, pressure changes, and a multitude of chemical compounds that can affect these chemical equilibria.

The take-home message from these studies is that biological complexity may arise from relatively simple systems that either function in a nonlinear way (as posited by Lanzara), or by the repetition of relatively simple steps (as posited by Mandelbrot and Wolfram), or by a combination of both. Endless complexities may be generated in this way, giving rise to complex adaptive

---

34 *Wikipedia*, s.v. "Complex systems biology," https://en.wikipedia.org/wiki/Complex_systems_biology.

systems that display feedback regulation and provide potential mechanisms for adaptability in the response to environmental stresses. We've only begun to explore these complexities.

Although this book suggests that we're formed from emergent chemistries, it is the magic and wonder of life and evolution that predict many more emergent phenomena. This book also suggests that there are many more connections and complexities to discover and understand. Just as the astronomers and physicists have discovered that the universe is far more complicated than they had originally imagined, so too havethose trying to define the complexities of life and evolution. If we've stimulated your mind to think of life and evolution in new ways, then we've succeeded in a major area of our joint endeavor. This enormous complexity shouldn't prevent us from trying to comprehend some small fraction of it.

CHAPTER 8

Fateful Encounters

Illustration by Gregory Lanzara (https://gregorylanzara.myportfolio.com/projects)

"You have to trust in something—your gut, destiny, life, karma, whatever. This approach has never let me down, and it has made all the difference in my life."

—Steve Jobs

HOW DO FATEFUL ENCOUNTERS INFLUENCE LIFE?
Fateful Encounters turn our lives around or upside-down. We're forced to adapt to these life-altering events. Going to war, falling in love, writing a book,

sex, and getting married are all examples of Fateful Encounters. The evolution of sex entails the fundamental idea of embedding Fateful Encounters into an organism's structures and functions. These are an overlooked but necessary part of all our lives and built into life's core.

We were born curious. This has made us all scientists, to some degree. We're also explorers traveling through life on our journeys of personal discovery. Writing this book has been a Fateful Encounter for me, requiring me to sum up Michael's and my discussions that, at the time, felt more like miniature vacations where we explored exotic places each time we talked. We had much fun exploring these exciting paths, which led to new growth and understanding that might have otherwise gone missing. Trying to capture that in this book has been challenging. This has also required new growth on my part.

Michael attempted to add new explorations to life's magnificence. He wanted to add the magical properties of life that suggest the increased potential for Fateful Encounters and Novelty. These added dimensions suggest that life has always found a way forward. If we question what this means for the chemical origin of life, we find that a chemical reaction might use slightly different chemical reactants to achieve slightly different chemical products. This type of reaction was most likely a catalytic reaction that expanded its chemical space into Novelty. The concept of creating an expanded chemical space prompts us to consider many basic reactions that led to a richer chemical space that might have evolved toward a metabolism. Fateful Encounters also include cooperation between molecules (i.e., modulation, cofactors, etc.), thereby creating expanded chemical interactions.

Fateful Encounters can enrich the biochemistries of life, which, in turn, can increase the overall health of an organism. Having options to make choices is my primary definition of overall health. The more options, the greater degree of health. Those who can choose to get out of bed have more options than those confined to bed. Similarly, those who can choose where to live because they have enough financial resources have better overall health than those who don't have such resources.

Fateful Encounters may also cause harm to the Self, which may not be lethal, but lead to cooperation or competition. They may also produce mergers, or symbiosis (change by interdependence or union, such as eating or being eaten). Fateful Encounters most likely merged different early cell types that led to the formation of the mitochondria and chloroplasts as organelles within a new type of cell we call a eukaryote. The cells of all the plants and animals on Earth today are eukaryotes.

Some say that our lives are controlled by fate; others say that we have choices at critical moments in our lives. The writing of this book is very much a conscious choice that I made, but I also realize that there were factors beyond my control that shaped the writing and publication of this book. Fateful Encounters not only allow us to counter fatalism with choice but present us with the opportunity of a life-altering encounter.

Mass Extinction Events

These events were the most dramatic Fateful Encounters that led to the formation of many new species. Since the beginning of life there have been five known mass extinction events that eliminated about 70 percent of all species. Interspersed among these periods were more minor extinction events. This suggests there must have been wild swings in the ability of life to survive for millions of years as a single species. The life forms that we know survived the longest as mostly unchanged organisms are the stromatolite-forming cyanobacteria.[1]

Whatever we do to life on Earth pales in comparison with the major extinction events of the past. That's the good news. Perhaps only a nuclear winter brought about by a full-scale nuclear exchange could stress life as much as the major extinction events of the past. That's the bad news. Life's chemistries appear to have been hardy enough to survive whatever has happened over the four-plus billion years on Earth.

Fateful Encounters produce scenarios of survival, dormancy, or death. Environmental changes such as earthquakes, floods, climate change, and volcanic eruptions are just a few of the environmental upheavals that may produce Fateful Encounters. The survival option for life often requires new features derived from previous chemistries. These may be a merger of two

---

1   *Wikipedia*, s.v. "Stromatolite," https://en.wikipedia.org/wiki/Stromatolite.

otherwise separate Selves, whereby their chemistries mutually benefit each other. An early example of a Fateful Encounter is when organisms drifted into extreme environments, such as volcanic hot springs, and had to adapt or die. The chemical interactions of life with its many environments has been the general arch of evolution for billions of years. We're just beginning to learn about these interactions and their impact on evolution.

Organisms that fail to adapt to their environments may eventually cease to exist. In order to adapt fully to environmental changes, organisms must have mechanisms that can pass adaptive changes from one generation to the next. Therefore, we must learn how life receives feedback from these changes, which are necessary for survival, and how it stores these changes as inheritable material. Fateful Encounters may have introduced the early protocells to potential memory molecules such as catalytic RNA. As mentioned, there is more than just the DNA in our genes that can function as inheritable material. Which molecular pathways might provide feedback to an organism's DNA to change genetic heredity? There are many ways to change an organism's DNA besides mutations. Mutations alone don't function as a direct feedback loop from the environment of an organism to its DNA because they are random; however, silencing DNA is one mechanism to control an organism's gene expression that may be a feedback from environmental changes. Gene silencing describes the epigenetic regulation of genes by other molecules such as proteins and RNA.[2] These are mechanisms from the proteins and enzymes within a cell's cytoplasm feeding back to the genome of an organism, which raises the question of which types of gene silencing are inherited.[3]

---

2   *Wikipedia*, s.v. "Gene silencing," https://en.wikipedia.org/wiki/Gene_silencing And also, Nature search - https://www.nature.com/subjects/gene-silencing.

3   Xuezhu Feng and Shouhong Guang, "Small RNAs, RNAi and the Inheritance of Gene Silencing in *Caenorhabditis elegans," Journal of Genetics and Genomics* 40, no. 4 (April 2013): 153–160, https://doi.org/10.1016/j.jgg.2012.12.007,   https://www.ncbi.nlm.nih.gov/pubmed/23618398. And also Mariusz Nowacki and Laura F. Landweber, "Epigenetic Inheritance in Ciliates," *Current Opinion in Microbiology* 12, no. 6 (December 2009): 638–643, https://www.ncbi.nlm.nih.gov/pmc/articles/PMC2868311/. See also, in answer to Michael's use-it-or-lose-it statement, Sophie Juliane Veigle, "Use/Disuse Paradigms Are Ubiquitous Concepts in Characterizing the Process Of Inheritance," *RNA Biology* 14, no. 12 (December 2, 2017): 1700–1704, https://www.ncbi.nlm.nih.gov/pubmed/28816621; Eugene V. Koonin and Yuri I. Wolf, Is Evolution Darwinian or/and Lamarckian?" *Biology Direct* 4, no. 42 (November 2009): 1–14, doi:10.1186/1745-6150-4-42, https://www.ncbi.nlm.nih.gov/pmc/articles/PMC2781790/

FATEFUL ENCOUNTERS MAY LEAD TO NEW GROWTH

All life naturally positions itself to engage in Fateful Encounters. They are a leap into new realities that might replace life's normal routines (homeostasis) and threaten life's very existence, but they also provide life with many rich and emergent chances at successful chemistries. They are also major structural or ecological reorganizations of life. They are the finite but incalculable probabilities that happen to all life. They are those life-altering events that shake and shape our lives such as man's discovery of fire. All life revolves around Fateful Encounters. Life is predisposed to experience Fateful Encounters and all of life's past Fateful Encounters are embedded within the chemistries of all current life. Once these Fateful Encounters happen, new things emerge.

Most stars are between one billion and ten billion years old, which means that life's current estimated age is on a par with that of the average star. Yet life had to have received the chemical remnants of at least one supernova, which exploded more than five billion years ago to give life its necessary elements (see Appendix II A). We may never truly know the details of those Fateful Encounters, which produced the chemical elements that began life almost four billion years ago.

The six major elements associated with all life—carbon, hydrogen, oxygen, nitrogen, phosphorous and sulfur (often abbreviated as CHONPS)—are the most important chemical elements that form the covalent combinations comprising most biological molecules on Earth. These chemical elements come from a wide variety of sources in the universe that include everything from the initial creation (the big bang created hydrogen) to the small and large stars (for the other chemical elements), including exploding white dwarf stars, which produce sulfur. This suggests a wide sampling of possible sources of the elements in the universe. We are made from chemical elements that were formed from supernova stars exploding many millions of miles away! If the coming together of these various and necessary chemical elements isn't an example of Fateful Encounters, then I don't know what is!

In a supernova explosion, atomic nuclei are bombarded with neutrons, making new chemical elements all the way up to uranium. These chemical elements are spread by the immense force of the supernova, which changes the composition of the gases, from which new stars can be reborn. Without supernova explosions, there would be no silicon to form rocky planets, no oxygen to form water—none of the chemical elements on which we depend here on Earth. This is the chemical stuff of which we are made. The nitrogen in our proteins, enzymes, and DNA, the calcium in our bones, the iron in our blood cells (hemoglobin), and the carbon in our proteins were all made in the interiors of stars. The silver and gold in jewelry and the silicon and copper in computers were all produced during stellar explosions. We ourselves—our entire world—are all made from these chemical elements, which are the byproducts of the cycles of many stars' birth and collapse. Truly a miraculous thought!

In biological terms, some of the earliest Fateful Encounters involved forming a membrane and growing as a protocell. Fateful Encounters brought these specific chemical elements together under the right conditions to create a new emergent protocell and begin the evolution toward life. Even if we eventually succeed in creating and manipulating life from basic chemical elements, we would not be able to say which events led to the creation of life on Earth. There are too many potential probabilities for us to pick the right ones. Finding life elsewhere in the universe may help us to understand it better here on Earth.

In the microbial world, swapping DNA and genes and engulfing other organisms are the Fateful Encounters among the various archaea and bacteria that may have led to the first eukaryote cell over two billion years ago. Fateful Encounters must have led these first eukaryote cells to begin colonizing together to form the first primitive, multicellular organisms such as sponges and jellies. This began a two-billion-year journey through many Fateful Encounters and evolutionary cycles to the multicellular organisms we see today. Eventually, this led to the dualism of Novelty and Stability that life requires (see Appendix I B). More than that, the road to life must have included the Essential Tensions that led to the development of Energy Flows and sensory proteins within the membrane. Having these basic components within an enclosed semipermeable membrane to create and direct Energy Flows and receive sensory inputs set the stage for chemical metabolisms that sense and respond to environmental conditions. Still lacking were the proteins (enzymes) necessary to create new membranes and the co-opting of the initial metabolism to make the additional proteins and enzymes necessary to support this early protocell. In these early years, Growth had to more directly contend with the balance between Stability and Novelty. Novelty took over when the growing protocell had to determine what to do with this Growth. Novelty led to the storing of energy and to the maintenance and division of the membrane. Novelty and the Essential Tensions paved the way toward cellular metabolism and homeostasis.

The tremendous complexities of the metabolic systems within living cells were the culmination of billions of years of evolution. They're the most ancient remnants of life's beginning. It is interesting that we exist at a time in evolution when chemical pollution is altering many organisms' metabolisms and biochemistries. Often, these changes are detrimental to the organism's Self, but some may prove beneficial. Eventually, life adapts to environmental changes and evolves to incorporate a new chemical landscape within life's biochemistries.

Earliest life must have had the most rudimentary tools for storing hereditary information. These earliest hereditary systems may have relied on template modelling using the already existing proteins that provided the Self with Stability. This may have also included catalytic forms of RNA molecules that could also function within the cell's metabolism (Ribozymes (ribonucleic acid enzymes) are RNA molecules that are capable of catalyzing specific biochemical reactions similar to the action of protein enzymes). Some of these sources of heredity may surprise you, but they often determine the direction of our lives.

Initially, it may have been both necessary and sufficient for the first life forms to use their membranes and their internal proteins and enzymes to convey the necessary structural memories for their complete repair and duplication. From the perspective of enclosed chemical species in a membrane, all these internal proteins and enzymes are transferred to a new Self or Selves. In the case of a mother and two daughter cells, the two daughter cells acquire roughly 79% old membrane from their mother cell (Appendix V: The Extra Membrane Necessary to Cover Two Daughter Cells Compared to the Original Mother Cell). Parts of this new enclosed membrane came from the older membrane of the original mother cell. The daughter cells also inherit 100% of the mother cell's cytoplasm. These are two often overlooked sources of hereditary information that can be passed from the mother cell to the daughter cells. The most important reasons for having heredity is for the maintenance of these underlying basic concepts to maintain the Self: Essential Tensions, Growth and

Energy Flow and Stability versus Novelty. Keeping the Essential Tensions is crucial for the sensory molecules (i.e. receptors) and Energy Flows are the secret sauce of life. As the criteria for Growth and Energy Flow, and Stability versus Novelty became larger in size and more complex in the number of networked chemical equilibria, and therefore, difficult to store as relatively simple templates, those molecules that could assume the roles as information molecules became ripe for emergence as the repositories of cellular and protein information. This suggests that there was a progression of changes that occurred in the earliest life that didn't have the full complement of hereditary and information molecules that we see today.

So much of life takes part in the very dark and messy places. -This is where you'll find sex. Fateful Encounters led to our conception, when the sperm penetrated the egg. It's impossible to know all the probabilities and circumstances that led up to this moment, yet we take it for granted that we were conceived and born. We may learn all that's necessary to do this artificially, but it is still a great mystery. After conception, our zygote (fertilized egg) had to be transported through the fallopian tubes to be implanted in the uterus of our mother. Thus, we've all made it through a long list of Fateful Encounters to be here in this moment in time.

DIFFERENCES BETWEEN NOVELTY AND FATEFUL ENCOUNTERS

Novelty builds on our existing Stability, whereas, Fateful Encounters create new realities that replace our normal routines and may threaten life's existence. Novelty and Fateful Encounters are two important components of life that ensure it can evolve, but the changes produced by Novelty and/or Fateful Encounters must be embedded within inheritable matter. At some point in the evolution of life on Earth one group began subsisting on light and carbon dioxide while another group subsisted on other sources of energy (food). This eventually led to a great divide between plants and animals. If we imagine the scenarios likely to have led to the beginning of animals, we might imagine Fateful Encounters that led to either cooperation or competition. Cooperation may have led to group behaviors while competition may have led to eating or being eaten. This was the beginning of consuming our fellow creatures as food. Organisms began eating each other because they were a good source of energy and because they could eat them without being eaten.

There's a dynamic balance between the restrictions to the possible chemical space and the expansions afforded chemical space (Stability versus Novelty and Fateful Encounters). In these evolutionary scenarios, Stability is a relative term. What we think of as stable systems may become unstable, given the impact of future events. After a catastrophic event such as an asteroid collision, if the Earth survives, life probably will survive too, but it will most likely be greatly changed. Novelty doesn't threaten immediate destruction. Novelty may exist quietly in the background and only be found useful during periods of stress, such as something new that broadens our skills and life.

Recruitment and Symbiosis

Quintessence suggests that life finds ways to incorporate Novelty into its biological repertoire. Cells in colonies will chemically attract and repel each other through chemical communication from escaped chemicals or wastes that can serve as messengers between individual cells. This creates the emergence of colony interactions that could lead to group synchrony, or recruitment, or subgroup symbiosis of chemical reactions that may provide benefits of an increased Energy Flow to all the cells involved.

This is an age of discovery. We're gradually understanding that we aren't what we thought we were. Thanks in large part to Ed Yong's fantastic book, *I Contain Multitudes*,[4] we now understand that we're made from whole societies of microbes and cells linked together in heretofore unbelievable ways! When we factor in our ignorance about most of the microscopic world, which includes the roles played by viruses, rogue proteins, plasmids, RNAs, DNAs and membrane transport vesicles between cells, we see that we're just beginning to understand this vast, mostly unexplored, macrocosm contained within the unseen microcosmic world.

---

4   Ed Yong, *I Contain Multitudes* (New York: HarperCollins, 2016).

CHAPTER 9

The Continuum

"Demasiada cordura puede ser la peor de las locuras, ver la vida como
es y no como debería de ser."

(Too much sanity may be madness. And maddest of all, to see life as it
is and not as it should be.)

—Miguel de Cervantes Saavedra

Like a surfer catching a wave, we're all perched on the cusp of an unfolding creation in a constantly emerging state. We have emerged as a unique set of chemistries that have evolved to sense, respond, and adapt. Were we formed by the hand of the Lord from the Earth, or did we evolve from the elements of the primordial universe? Either way, it is truly a miracle!

Scientifically, we know that we're made from star stuff. Our hemoglobin molecules that carry our life-sustaining oxygen in our red blood cells contain iron, which was formed from exploding supernova stars in outer space. In religious terms, we may believe we evolve into higher spiritual beings through a mysterious spiritual grace. When we see the scientific and religious concepts together, they give us an awe-inspiring view of the universe. This is the basis for those in science who are believers and those in religion who are scientists. Those who require the tyranny of fixed ideas see little of the miraculous!

Our unique set of chemistries has brought us to this point in time where we can peer over the edge of creation and contemplate our existence. Life's search for Novelty (Neosis) and subsequent attempts to embrace and incorporate it (Quintessence) have created multiple new paths that have emerged and will continue to emerge with surprising results. It was life's earliest chemistry that captured Novelty and transformed it into Stability, which was accomplished by linking together networks of chemical equilibria into systems that could expand and grow. These linkages stabilized and formed biochemical networks, which increased potential branching chemistries. Eventually, this led to the dualism of Novelty and Stability that life requires. Fascinating as this is, these relatively simple chemical equilibria still exist today in all our sensory receptor molecules. How organisms began sensing the environment is perhaps one of the most important and often overlooked areas concerning the emergence of life. Every aspect of our bodies from our senses to our genes is regulated by chemicals and the underlying principles of chemistry. This book has discussed only a small fraction of those beautiful and disparate chemistries that compose life and evolution.

At this point in time, it seems inconceivable that science may someday explain all possible chemistries that life and evolution might encounter, including those of Novelties and Fateful Encounters. What lies hidden beneath these processes are the very basic biochemical processes that also allow for an already-existing molecule, such as a cellular enzyme, to acquire a new function, thereby capturing Novelty. These processes have happened many times in the past and are intimately involved in capturing Novelty and transforming it into Stability. Currently, we've extended our sensory receptors by the introduction of the Internet, which is perhaps a good example of capturing Novelty and transforming it into a new form of relative Stability. Now we take for granted that the Internet is there and ready to use for quick reference, business, or social media. This has greatly extended the reach of our minds into the world. Many of us have grown up in this new age without giving much thought to the consequences, but whatever the consequences, we've embraced Novelty wholeheartedly and embarked on a wonderful new era of exploration.

In this book, we've presented how many scientists describe life and evolution. We've also shown that science has a long way to go. Life has continuously evolved in an unbroken chain, in which all life can trace itself back to that beginning. This also suggests that all life on Earth is related to every other life in many interesting and surprising ways. Since life's beginning on Earth, there's been an explosion of life that appears to fill every available niche, which also represents an enormous explosion in the chemical diversity of life. Life functions much like a chemical catalysis that encounters and interacts with different environments. A plethora of new chemical combinations have been created since the existence of life, which include natural toxins as well as the millions of new chemical compounds that are considered unnatural pollutants created by us humans. These new chemicals that have been introduced into the environment increase the possibilities for new chemical combinations with future life for ages to come. This may alter the arc of evolution to include new life forms that we might not expect.

There can be no doubt that Michael's goal was to expand the reader's view of life and evolution. This expanded view is an important theme of this book, which suggests that being alive creates new opportunities for emergent phenomena that we couldn't have previously recognized. One of Michael's insights was to take a more adaptive approach to life and evolution. One of the central themes in this book is that evolution depends on the adaptive chemistries found in all life. As we learn more about life, how will we cope with this knowledge? It is a vitally important question. Some themes that emerge are Michael's desire to emphasize life's magic, Fateful Encounters, search for

Novelty, and drama. In his opinion, cell membranes, metabolism, growth, and reproduction are four qualities often associated with being alive, and they are also included within our five essential concepts defining life. Michael's evolutionary dream was to show how living organisms capture Novelty and transform it into Stability. He also wanted to show that Quintessence offers a new explanation for genetic Novelty and its role in evolution by explaining how life captures the Novelty of chance encounters, negotiates through opposing pressures, and grows and evolves by internalizing interactions. On a chemical level, Quintessence is the search for, and capture of, Novelty (Noesis), which is eventually turned into Stability (homeostasis). From a broader perspective, Quintessence represents life's essential balance between Stability, Novelty and Fateful Encounters.

Growth and genetic Novelty involve biochemical interactions of an organism's metabolism with genes and the epigenetics of how those genes are controlled. Michael considered various ideas in exploring this puzzle. He thought there might be mechanisms, as yet undiscovered, that could transfer epigenetic information from an organism to its genes. Some of the mechanisms he considered were retroviral transfer from rogue gene elements, which included horizontal gene transfer, and epigenetic modifications of gene function and expression. He and I were stymied by the fact that we knew organisms could change to better suit their environmental conditions, but many of these mechanisms weren't well studied or accepted as inheritable factors. Currently, there's a rich and burgeoning chemistry of biochemical heredity that encompasses epigenetic and genetic phenomena that we haven't fully explored in this book.

Humans are perhaps the first species to have dominion over all other species on Earth. We're relatively young as a species. We might be the beginning of a grand continuum of beings that will gradually acquire greater and greater powers over the physical universe and our own lives. From our perspective, the grand ark of life appears to be in a transformative period where we gain greater control over our own biology and the biology of other life forms. This may begin the greatest leap of exponential growth in life's future development. Soon we may begin creating new species and Selves. These possibilities foreshadow our previous development for which we are mostly unprepared. This book may help prepare us by opening our eyes to the potentials of future outcomes.

Someday we might have a full molecular description of life. This should in no way detract from it. We've discovered several of the hidden secrets of the atom, yet there exist many more mysteries to explore. So too, with life. We will never exhaust all mystery in the universe, which includes our ability, or lack thereof, to comprehend it. We crave insights into that beauty and mystery. As Richard Feynman suggested, science adds to the overall mystery rather than detracts. Science explores these mysterious elements of life. For every question science answers many more questions arise. We should always question, because it is inherent within our foundational chemistries.

Quintessence extends our identity from a species that just lives to survive, compete, and reproduce, to a species that also lives for stimulation, connection, and creation. If we accept this new identity, it changes not only how we understand living with each other but also how we understand our relationship with all life. In the end, Quintessence helps ground us on how we got here and opens our imagination of where the future course of evolution in biology, culture, and technology may take us.

Our small attempts at understanding only demonstrate how inadequate we are. In the end, the wisdom of life lies not in understanding it but, rather, in how much we have loved. As T. S. Eliot put it in his poem "Little Gidding": "*We shall not cease from exploration*, and the end of all our *exploring* will be to arrive where *we* started and know the place for the first time." Michael and I were both on this journey of questioning and exploring. We grew. I was lucky enough to catch the wave with him.

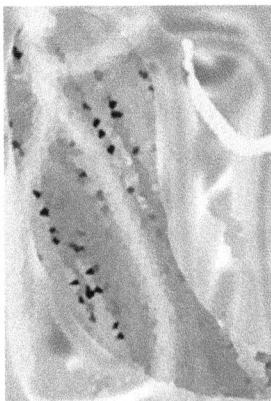

Glossary

Anabolic: The metabolic reactions that construct the molecules of living organisms (the Self).

Atom: The smallest particle that retains the chemical properties of a specific chemical element.

Autocatalytic: The reactions that catalyze (speed up) themselves.

Autoinhibition: Describes an enzyme (or receptor) action that decreases or ceases its response after repeated or prolonged presentations of a substrate (or stimulus). Other terms are habituation, tolerance, hormesis, desensitization, fade, tachyphylaxis, wearing-off, down-regulation, biphasic dose-response, J-shaped dose-response.

Bacterial conjugation: The transfer of genetic material between bacterial cells by direct cell-to-cell contact or by a bridge-like connection between two cells.

Catabolic: The metabolic reactions that tear down the molecules of living organisms (the Self).

Chemical tensions: Chemical equilibria usually separated by a membrane and comprising at least two chemicals linked together by a process that can change one form into the other.

Chemiosmotic coupling (chemiosmosis): The movement of chemical ions across a semipermeable membrane and down their electrochemical gradient.

Desensitization: A decrease in response (or effect) with a larger or greater stimulus.

Disequilibrium: A perturbed equilibrium.

Essential Tensions: Those chemical tensions necessary for life (the Self).

Eukaryote: Living organisms that have a nucleus enclosed in membranes. (See https://en.wikipedia.org/wiki/Eukaryote.)

Fateful Encounters: Encounters that dramatically change the environmental biochemistries surrounding the Self.

Gradient: A change in an amount of something such as a chemical (e.g., an odor) over a range of space.

Habituation: A form of learning in which an organism decreases or ceases its responses to a stimulus after repeated or prolonged presentations.

Homeostasis: The tendency of a system, especially the physiological system of higher animals, to maintain internal stability, owing to the coordinated response of its parts to any situation or stimulus that would tend to disturb its normal condition or function.

Ion: A charge that can be either positive or negative on a chemical element or molecule.

Ionic gradient: A gradient of electrochemical potential for ions that can move across a membrane. (See https://en.wikipedia.org/wiki/Electrochemical_gradient.)

Metabolic pathway: A linked series of chemical reactions occurring within a cell. (See: https://en.wikipedia.org/wiki/Metabolic_pathway)

Molecule: Any composition of matter comprising two or more atoms.

Negative feedback: A reaction that causes a decrease in function, which occurs in response to a stimulus.

Net shift: The perturbation produced by either a molecule or force interacting unequally with the two sides of a chemical equilibrium.

Non-Essential Tensions: non-Essential Tensions are those disequilibria that exist but are not essential tensions for life.

Novelty: Something outside the Self and its extended environment(s).

Point charge: A charge, either positive or negative, in a specific chemical element (atom) within a molecule.

Protist: Any eukaryotic organism (one with cells containing a nucleus) that is not an animal, plant, or fungus.

Protocells: Membrane-bound chemical entities that may have existed before the formation of cellular life.

Receptors: Molecules that can sense changes in their environment and transmit that change (signal) to other molecules within living organisms.

Redox: Reduction and oxidation reactions transfer an electron to or from chemical elements or molecules.

Redox disequilibria: Any redox reaction can be broken down into two half reactions: oxidation at the anode (loss of electrons) and reduction at the cathode (gain of electrons). In the presence of a semipermeable membrane, the normal equilibrium may be perturbed to create a disequilibrium.

Ribozyme: Ribozymes (ribonucleic acid enzymes) are RNA molecules that are capable of catalyzing specific biochemical reactions, similar to the action of protein enzymes. (See https://en.wikipedia.org/wiki/Ribozyme.)

Self: The smallest portion of a living organism that retains the complete biological properties of that organism (similar to the concept of an atom to the chemical elements).

Self-benefit: A process that leads to an increase in Stability, Growth and/or Energy Flow for a living organism (Self).

Stability: Anything that contributes to the perpetuation of the Self or a Species.

Stromatolite: Layered mounds, columns, and sheet-like sedimentary rocks that were originally formed by the growth of layer upon layer of cyanobacteria, a single-celled photosynthesizing microbe. (See https://en.wikipedia.org/wiki/Stromatolite.)

Symbiosis: A connection between different organisms, in which they live together and benefit from each other.

Two-state system: A two-state system is any chemical equilibrium between active and inactive states.

Tachyphylaxis: Rapid desensitization, often occurring in the microsecond to seconds time range.

Weber's law (also called the Weber-Fechner law): A just-noticeable difference in a stimulus needs to be proportional to the magnitude of the original stimulus in order to be sensed or noticed. This implies that with a larger original stimulus (such as a bright light) there must be a compensatorily larger difference between the two stimuli—such as the background bright light and the light from an object such as a flashlight—to notice (or perceive) a difference.

The practical explanation is that a flashlight won't be seen in the bright light of a search light because the light from the flashlight doesn't have enough brightness to be seen above the brightness of the search light.

Zygote: A eukaryotic cell formed by a fertilization event between two gametes. The zygote contains all of the genetic information necessary to form a new individual. In multicellular organisms, the zygote is the earliest developmental stage. In single-celled organisms, the zygote can divide asexually by mitosis to produce identical offspring, https://en.wikipedia.org/wiki/Zygote.

Appendixes

APPENDIX I A: THE CREATION IMPERATIVE
by Dr. Michael Kuperstein

Why are we here? If you say, evolution, that may answer the question of how we got here but not why. The answer to why we exist needs to include not only all of life's meaningful experiences but also the experiences of all of life that came before us, since we are the result of the evolution of prior species. The answer needs to include a grand continuity of life, to capture the range of experiences from the simplest life forms to the most complex personal, social, and cultural experiences we have in our lives. And as life forms get more complex, the grand continuity must be naturally extensible to meet whatever level of complexity life forms take.

As human beings, most of us get the meaning of our lives from the stories we live through, whether they be stories of redemption, what we built, what we discovered, who we served, who we belong to, how we won or lost, whom we loved, and what we hoped for. Our stories live inside us from cradle to grave.

But to get started, our stories need a foundation and a driving process. Our stories inherit a driving process that pushes both the unfolding of our lives and their continuous creation. Sure, we inherit the genes for how we look, where we were brought up, our heritage and how we were raised. But what do we inherit from our fetus before we are born, besides how we will look? What is it inside us that drives what we do, our will, and the stories that become the meaning of our lives?

To find out, let's start when our blueprint first formed: the joining of egg and sperm. The conceived egg cell initially multiplies into exact copies. Then these exact copies begin to differentiate in the first of thousands of ways to become a fetus that develops, in most cases, into a normal birth. The key to this important differentiation process is a competitive and cooperative process between groupings of those nearly exact cell copies. How they differentiate and migrate in the fetus depends on where the cell groups are located along a set of chemical gradients across the tiny fetus, and their position within these gradients signals the generation of specific proteins from our DNA that determines their fate as skin cells, heart cells, brain cells, and others.

As the fetus grows, this differentiation process creates increasingly complex layers of tissues and communication systems stage by stage until a fully formed human being can begin its life outside the womb. And with our first cry at birth, our stories begin.

What our stories inherit from the development of our fetus are internal and external biological forces that continuously compete and cooperate with each other throughout our lifetime. And these same forces are not only in effect during our lifetime but have been since biological life began almost four billion year ago. These biological forces have been a part of every species and every evolution from the first bacteria to us humans.

On a daily basis, we take these forces for granted because they are an unnoticed part of us, but understanding what they are and how and where they work has a profound influence on why we are here and the stories of our lives. We are here because these biological forces drive our growth as well as how our environment interacts with our internal forces to set the stage for evolution. Not only are these forces responsible for our birth process, which our life stories inherit, but they also continue to act on our stories throughout our lifetime.

To understand these forces, we need to go back to the essential properties for life to exist. All life forms are composed of cells. Each cell is enclosed by a membrane that distinguishes it from its environment and other cells. Each cell draws nourishment and energy from its environment that feeds the cell's metabolism, which self-regulates its existence. In addition, each cell needs to be reactive to both its changing environment and its internal state in order to adapt and survive, and each cell needs a mechanism to reproduce.

Let's focus first on how two of these properties interact. A self-regulating metabolism maintains the cell's internal stability, while the process of adapting to the cell's environment requires that the cell be able to prepare for, engage in, and evaluate novel interactions. In doing so, the cell can orient to or away from its changing environment to capture nutrients or escape danger. These two processes are in continual competition for the cell's energy since they are driving toward very different results, one to maintain Stability with its known benefits and the other to engage in novel interactions that have unknown future benefits. How much energy should a cell commit to Stability versus novelty? All Stability and no novelty will result in the cell's death due to a changing environment. All novelty and no Stability will also result in the cell's death because the cell will just fall apart, always seeking novel interactions.

These two processes need to remain in some dynamic competitive and cooperative balance where one or the other may dominate for some time and then they might reverse. The notion that opposing processes are maintained in a dynamic balance has been known for thousands of years as dualism, from Taoism's yin and yang. Many other examples in our own daily experience display dualism, including the experiences of owning versus sharing, of separating versus belonging, of competing versus cooperating, and on and on.

Claude Bernard, in 1885, was the first to study the way life maintains Stability in a process later called homeostasis. But no one has yet coined a name for the process of searching for novelty. So I will call it Neosis.[1] As I show examples of how the forces of homeostasis and Neosis act, you can see the beginning of a grand continuity between our biological being and our emotional and conscious being, which will get us back to our original question of why we are here.

The essential features for life to exist did not just appear out of nowhere. Those features are themselves inherited from the physics and chemistry of Earth in our oceans (Appendix II A). As Earth's plate tectonics shifted, the areas near their cracks often oozed lava from below the crust. When the hot lava met the cold ocean, something special happened at the boundary. All sorts of gradients started to form from adjacent differences in temperature, chemistry,

---

1   Neosis has been previously used to describe the uncontrolled growth of cancer cells. See https://www.ncbi.nlm.nih.gov/pubmed/16316756, and https://www.ncbi.nlm.nih.gov/pubmed/14726689.

acidity, ionic concentrations, and pressure, across this boundary. These gradients drove precipitation of chemicals in the rising water flows above the lava, as well as drove ionic flows across the boundary between the ocean and these rising flows. At a macro level, tall chimneys of chemical precipitates, called hydrothermal vents, were created. They look like huge underwater termite mounds spewing fast-rising chemical flows from the cracks in the earth.

For our discussion, the most important events took place on a microscale. The surface of the chimney-like structure of the hydrothermal vents were dimpled with tiny pockets, the size of today's cells. These pockets were lined with sulfur, iron, and organic chemical compounds that came from the lava's chemistry. Some of these compounds came as chemical strings and sheets, and of these, some self-assembled from standard chemical reactions to line the tiny pockets of the vent surface and create protocells, which are nonliving, organic chemicals encased in membranes resembling today's cells.[2]

In addition, ionic gradients across the ocean-vent boundary, created flows of ions that permeated these protocells across their membranes, thereby injecting them with energy, known as the first proton pumps. These pumps enabled the first enzymes inside the protocells to bond with protons and electrons, which turned on organic chemical reactions that led to the first metabolism, and eventually, to the first cell divisions and genetic reproductions, as life began.

One main point here is that when life began, the means for energy flow started around the hydrothermal vents, which eventually evolved into cell digestion and cell breathing and this energy flow fueled the first self-regulating metabolism that converted nourishment into cellular structures that maintained the cell's Stability. The second point is that in order to survive in a changing environment, the cell's membrane was required to respond and

2   William Martin and Michael J. Russell, "On the Origin of Biochemistry at an Alkaline Hydrothermal Vent," *Philosophical Transactions of the Royal Society B: Biological Sciences* 362 (November 3, 2006): 1887–1926, doi:10.1098/rstb.2006.1881; Michael J. Russell et al., "The Drive to Life on Wet and Icy Worlds," *Astrobiology* 14, no. 4 (April 15, 2014): 308–343, doi:10.1089/ast.2013.1110; Wolfgang Nitschke and Michael J. Russell, "Beating the Acetyl Coenzyme A-Pathway to the Origin of Life," *Philosophical Transactions of the Royal Society B: Biological Sciences* 368 (July 19, 2013): 1–15, doi:10.1098/rstb.2012.0258.

adapt to novel aspects of its environment. And so began the forces for seeking novelty and maintaining Stability fueled by the flow of physical energy in the form of ionic flows.

We have come a long way from where we started. We went from life stories to what we inherit from the developing fetus to what we inherit from the earliest life forms to the origin of life on Earth. Where am I going with this? I am beginning to show the outline of the grand continuity of life forces and how they impact the meaning of our lives.

APPENDIX I B: FIVE ESSENTIAL CONCEPTS DEFINING LIFE

The five essential concepts defining life (or criteria for life) conceptualize the overall balance of dichotomies across life and evolutionary history. As we delve into each of these concepts, we will begin to appreciate the structure they provide for our path forward.

1. The *Self versus the non-Self.* By defining the Self, we set it in opposition to the non-Self. Self and non-Self are separated by a type of barrier, such as a semipermeable membrane. This definition of Self also requires the presence of the other four concepts listed below.

2. *Essential Tensions versus non-Essential Tensions* (see Glossary for a definition of non-Essential Tensions). Sets of chemical systems are coupled together in ways that prevent any of the them from realizing their normal chemical equilibrium. Such imbalances most likely exist as structural imbalances due to the nature of the semipermeable membrane. A semipermeable membrane has unique properties that allow some chemical species to penetrate the membrane, while others are not able to do so. Such poised chemical imbalances are essential for all life, as sources of energy for metabolism, and they also include the sensory systems. At the most primitive level these sensory systems must couple with other systems such as the ion permeability of the membrane or phospholipid production.

3. *Growth and Energy Flow versus stasis (dormancy) and no Energy Flow.* This is the dichotomy that encompasses the use and storage of extra molecules and/or energy. It is the process whereby the proteins and enzymes of various metabolic cycles control and direct energy into molecules that store energy or new proteins and enzymes. Growth occurs when the Energy Flow contributes to the creation of the structural and necessary metabolic molecules of the Self. This is also referred to as anabolic metabolism. (Stasis is the same as being dormant and no Energy Flow is when there is no energy source to support life. With no Energy Flow life either becomes dormant or dies.)

4. *Stability versus Novelty.* The dynamic balance between the Stability-enforcing processes and molecules of the Self in opposition to the novel expansion of potential physical and chemical spaces such as new energy sources or chemical functions. There exists a restricted chemical space that defines the overall usefulness of any new proteins or molecules. This potential chemical space of the Self may be expanded by novel encounters with new molecules or by creating new proteins/molecules from existing versions, such as by chemical modifications. This may facilitate important modifications, among which are the chemical modifications made by various secondary processes within cells, such as methylation, phosphorylation, and acetylation. The interplay of Stability with Novelty produced mechanisms that eventually emerged as inheritance.

5. *Fateful Encounters versus no Fateful Encounters.* Fateful Encounters allow for new chemistry to link with an organism's existing metabolism and biochemistry. Often, these changes are detrimental to the Self, but some may prove beneficial. Fateful Encounters can enrich the biochemistries of life. Depending on the type and number of environments the Self occupies, such encounters may lead to either cooperation or competition. They may also lead to more extreme outcomes. These include symbiosis, (which produces change through interdependence or union), and competition (which can lead, for example, to eating or being eaten). Fateful Encounters may also happen within the Self, potentially leading to Novelty, and eventually, Stability as mentioned in the fourth concept above. No Fateful Encounters are those changes that are either not fateful or don't happen. They don't propagate a compensatory change in the organism or just fail to happen.

We can also apply these five essential concepts for life to evolution.

1. *Self versus non-Self.* In evolution, this refers to the level of a species.
2. *Essential Tensions versus non-Essential Tensions.* These are those sets of imbalances that separate species from each other.

3. *Growth and Energy Flow versus stasis (dormancy) and no Energy Flow.* Growth is required to maintain a species and/or evolve new species. This Growth depends upon the Energy Flow (food). If the Energy Flow is reduced or disrupted, Growth may not occur if alternative resources aren't available to sustain it. This may lead to either death or stasis as dormancy.

4. *Stability versus Novelty.* The dynamic balance between the Stability-enforcing mechanisms of a species in opposition to the novel expansion of physical and chemical spaces for potential new species.

5. *Fateful Encounters versus no Fateful Encounters.* – These are the potential life-altering effects resulting from the potential physical, chemical, and biological encounters in the life of a group of organisms or species that may have life-altering effects. Fateful Encounters depend on the type of environment that a species occupies. They can also happen within a species, which may lead to Novelty, and eventually, to Stability. Some of these encounters may cause harm or extinction to a species. Others may place nonlethal stress on a species. They may also lead to either cooperation or competition among interacting species.

REFERENCES

These references, while far from extensive, cite indirectly or directly the five concepts discussed above.

Emmeche, Claus. "Defining Life as a Semiotic Phenomenon." *Cybernetics and Human Knowing* 5, no. 1 (1998): 3–17, http://www.nbi.dk/~emmeche/cePubl/97e.defLife.v3f.html.

Ho, Mae-Wan, and Robert Ulanowicz. "Sustainable Systems as Organisms?" *BioSystems* 82, no. 1 (October 2005): 39–51.

Koshland, D. E. Jr. "The Seven Pillars of Life." *Science* 295, no. 5563 (March 2002): 2215–2216.

Kuperstein, Michael, and Terry Deacon. "Workshop on Self-Determination in Developing and Evolving Systems." Harvard University, January 6–9, 1994.

Mayr, E., *The Growth of Biological Thought: Diversity, Evolution, and Inheritance.* Cambridge, MA: Belknap Press, 1982.

———. "The Ontological Status of Species." *Biology and Philosophy* 2, no. 2 (April 1987): 145–166. Reprinted in Mayr, *Toward a New Philosophy of Biology.* Cambridge, MA: Harvard University Press, 1988.

Patten, B. C., M. Straskraba, and S. E. Jørgensen. "Ecosystems Emerging: 1. Conservation." *Ecological Modelling* 96, nos. 1–3 (March 1, 1997): 221–284.

Smith, Maynard J. *The Problems of Biology.* Oxford: Oxford University Press, 1986.

———. "Evolution: Natural and Artificial." In *The Philosophy of Artificial Life.* Edited by M. A. Boden. Oxford: Oxford University Press, 1996, 173–178.

Zhuravlev, Y. N., and V. A. Avetisov. "The Definition of Life in the Context of Its Origin." *Biogeosciences* 3 (July 10, 2006): 281–291, https://doi.org/10.5194/bg-3-281-2006.

APPENDIX II: THE CHEMICAL ELEMENTS OF LIFE

This table lists the more common essential chemical elements and where they're found in the universe.

The Chemical Elements at Various Locations in the Universe
(Numbers in parentheses refer to References, below.)

Number Element number from the periodic table	Chemical Element (symbol, name)(1)	Rank of abundance in our galaxy (The Milky Way)(2a or 2b)	Rank of abundance in Earth's crust (3)	Rank of abundance in sea water (3)	Rank of abundance in the human body (4)
1, 1	H (Hydrogen)	1 or 1	10	2	3
2, 6	C (Carbon)	4 or 6	17	10	2
3, 8	O (Oxygen)	3 or 8	1	1	1
4, 7	N (Nitrogen)	7 or 7	34	15	4
5,15	P (Phosphorous)	-	11	19	6
6,16	S (Sulfur)	10 or 16	16	6	8
7,11	Na (Sodium)	-	6	4	9
8,19	K (Potassium)	11	8	8	7
9,20	Ca (Calcium)	-	5	7	5
10,12	Mg (Magnesium)	9 or 12	7	5	11
11,17	Cl (Chlorine)	-	19	3	10
12,26	Fe (Iron)	6 or 26	4	29	12
13,30	Zn (Zinc)	-	24	24	14
14,9	F (Fluorine)	-	13	14	13
15,25	Mn (Manganese)	-	12	36	32
16,27	Co (Cobalt)	-	30	49	39
17,29	Cu (Copper)	-	26	34	20
18,42	Mo (Molybdenum)	-	58	23	37
19,34	Se (Selenium)	-	68	37	29
20,53	I (Iodine)	-	62	20	26
21,14	Si (Silicone)	8 or 14	2	13	15
22,28	Ni (Nickel)	12	23	31	30
23,24	Cr (Chromium)	-	21	32	31
24,35	Br (Bromine)	-	50	9	18
25,23	V (Vanadium)	-	20	27	54

Major Elements: Carbon, hydrogen, oxygen and nitrogen are known as the main "organic" elements because they form the building blocks that make life possible. Among the four, carbon is perhaps the most special, since it can form bonds with itself and makes molecules that have many different shapes. Carbon molecules can be short chains, long chains, bent chains, branching chains, contain multiple bonds, form ring shapes among a plethora of possible combinations. The four classes of macromolecules that make life possible (protein, carbohydrates, lipids, and nucleic acids) are all made of carbon, along with the other three main organic elements. Aside from the big four mentioned above, the next major elements are phosphorus, sulfur, sodium, chlorine, potassium, calcium, and magnesium. Phosphorous may be an important limiting element when it comes to searching for alien Life. For life to begin, it may be necessary for only the first twelve elements in the table. Other research indicates that Ag and Cd sulfides may have been important for life's beginnings (5), Or an ancestral methanotrophic pathway (6).

About 19 of the approximately 118 known chemical elements are essential for humans. An essential element is one whose absence results in abnormal biological function or development that is prevented by dietary supplementation with that element. Living organisms contain relatively large amounts of oxygen, carbon, hydrogen, nitrogen, and sulfur (these five elements are known as the bulk elements), along with sodium, magnesium, potassium, calcium, chlorine, and phosphorus (these six elements are known as macrominerals). The other essential elements are the trace elements, which are present in very small quantities. Dietary intakes of elements range from deficient to optimum to toxic with increasing quantities; the optimum levels differ greatly among the essential elements.

REFERENCES

1.   Chellan P., and P. J. Sadler. "The elements of life and medicines." *Philosophical Transactions of the Royal Society A: Mathematical, Physical and Engineering Sciences* 373 (2015), http://dx.doi.org/10.1098/rsta.2014.0182.

2a. *Wikipedia*, "Chemical element," https://en.wikipedia.org/wiki/Chemical_element.

2 b. *Wikipedia*, "Abundance of the elements," https://en.wikipedia.org/wiki/
      Abundance_of_the_chemical_elements.

3.   *Wikipedia*, "Abundance of the elements (data page)," https://en.wikipedia.org/wiki/
      Abundances_of_the_elements_(data_page)

4.   *Wikipedia*, "Composition of the human body,"
      https://en.wikipedia.org/wiki/Composition_of_the_human_body.

5.   Kitadai, Norio et al. "Geoelectrochemical CO Production: Implications for the
      Autotrophic Origin of Life." *ScienceAdvances* 4, no. 4 (April 2018): eaao7265,
      doi:10.1126/sciadv.aao7265. https://www.ncbi.nlm.nih.gov/pmc/articles/
      PMC5884689/.

6.   Russell, Michael J. "Beating the Acetyl Coenzyme A-Pathway to the Origin of Life."
      Philosophical Transactions of the Royal Society B: Biological Sciences 368, no. 1622
      (July 19, 2013), https://www.ncbi.nlm.nih.gov/pmc/articles/PMC3685460 (M's
      favorite).

APPENDIX III A: ESSENTIAL TENSIONS

From basic chemistry, life inherited the Essential Tensions that are necessary for all life. In the complex interplay of these chemical phenomena, the Essential Tensions developed as perturbations in the normal chemical equilibria. Essential Tensions describe those characteristics of life that produced disequilibria, which minimally involves a two-state chemical reaction. These reactions are prevented from reaching their normal equilibrium due to the presence of a semipermeable membrane or other interfering molecules that separate the two sides of a chemical equilibrium. This prevents the equilibration that would occur naturally. Once these chemical disequilibria began, which was very early in the universe, the potential for the ignition of life's flame also began.

The coupling of these Essential Tensions to other molecules, including enzymes, or sensory molecules such as receptors, becomes a critical process on the road to life. We don't know enough about which types of couplings can create a living system, but we may be closer to understanding some aspects of life's essential chemical core.

Feedback control is an inherent mechanism in the Essential Tensions that is important for understanding the underlying scientific basis of sensory cognition, memory, and learning. Feedback control, which is at the heart of Wallace's Essay on the evolutionary principle[1] begins to unravel a marvelous mechanism that they all have in common: a simple balance, which is not so simple.

The main reason that a simple balance is not so simple is that the parameterization of the net shift of the equilibrium of a simple balance unexpectedly produces nonlinear dynamics, which are exactly the mathematics required to describe receptor and enzyme complexes in living organisms. This is explained in more detail in Appendix III B and C. The bottom line is that by using a derivation from the balance and an expression for chemical binding (Langmuir binding, which describes how one molecule binds to another molecule), most of the nonlinear dynamics of biological and sensory systems can be explained and accurately reproduced. This has major implications for biology and other areas of science as well.

The reason that feedback control is important to so many areas including evolutionary biology is that it works through a negative feedback loop that prevents the overall system from straying too far from an equilibrium point (homeostasis). The major question here is what in nature could provide these negative feedback loops to the biochemical reactions of living organisms? The answer is that they are provided by the chemical principles associated with a simple equilibrium balance. This relatively simple system embraces fundamental chemical observations made more than a century ago by Le Chatelier.[1]

Le Chatelier's principle, also called the equilibrium law, can be used to predict the effect of a change in conditions on chemical equilibria. It can be stated as follows:

> When any system at equilibrium for a long period of time is subjected to change in concentration, temperature, volume, or pressure, the system readjusts itself to partly counteract the effect of the applied change, and a new equilibrium is established (see Reference below).

In other words, whenever a system in equilibrium is disturbed, the system will adjust itself in such a way that the effect of the change will be reduced or moderated. This fundamental principle has a variety of names, depending upon the discipline using it—for example, *homeostasis*, a term commonly used in biology. It is common to state this as follows: any change in status quo prompts an opposing reaction in the responding system.

In chemistry, the principle is used to manipulate the outcomes of reversible reactions, often to increase the yield of reactions. In pharmacology, the binding of ligands to the receptor may shift the equilibrium according to Le Chatelier's principle, thereby explaining receptor activation and the subsequent negative feedback of rapid desensitization or tachyphylaxis.

It is this aspect of negative feedback control that is so fundamental to many important biological processes. It occurs when an overwhelming amount of a stimulus reacts with our sensory systems. It could be too much light, for example: we may not see clearly after exposure to intense light. This phenomenon happens with all our sensory systems. Overstimulation produces a temporary decrease in our perception. This feedback is a necessary component for all life. There are many scientific and technical terms for such negative feedback, which is confusing to the scientist and layman alike. It may, for example, be referred to as desensitization, habituation, bell-shaped curves, down-regulation, biphasic response, hormesis, autoinhibition, substrate inhibition, tachyphylaxis, wearing-off, fade, or tolerance. The effect on our sensory systems is that they become unresponsive to further stimuli (or substrate, in the case of enzymes). This is naturally hard-wired into the chemical responses of these systems.

The fact that Le Chatelier never quantified his principle has entailed much scientific discussion. In this book, we show a quantification of Le Chatelier's principle in the following, Appendix III B.

REFERENCE
Wikipedia, "Le Chatelier's principle," https://en.wikipedia.org/wiki/ Le_ Chatelier%27s_principle.

APPENDIX III B: THE QUANTIFICATION OF LE CHATELIER'S PRINCIPLE

L e Chatelier failed to quantify his statements about the effects of perturbations on the initial chemical equilibrium. By introducing a parameter for these perturbations, these effects can be characterized as a net shift in the underlying equilibrium. This net shift provides a fundamental understanding of many chemical and biological principles. Much of life's chemistry is governed by these perturbations in biological systems of chemical and enzymatic equilibria.

In general, given an equilibrium constant, Keq, representing the starting equilibrium conditions of reactants and products:

$$A \rightleftharpoons B$$
$$Keq = \frac{[B]}{[A]}$$

Any perturbation to Keq can be expressed as:

$$Keq = \frac{[B] + \Delta}{[A] - \Delta}$$

Where Δ represents the initial overall change to Keq obeying the conservation of mass law. Depending on the type of perturbation, Δ will return to zero, over some time period, to restore the original equilibrium. This is true only if the perturbation is transient and produces no permanent changes to the original equilibrium system.

As an example, if we add some relatively small amount to A, giving $A + x$, then we have the following:

$$Keq = \frac{[B]}{[A + x]} = \frac{[B] + \Delta}{[A] - \Delta}$$

And solving for Δ gives:

$$\Delta = \frac{-x[B]}{[A] + [B] + x}$$

Which is the initial perturbation to the system. The parameter Δ demonstrates how a perturbation to the original equilibrium constant, *Keq*, would occur by a net shift in the amounts of *A* and *B* alone—that is, if *x* were not added and the initial equilibrium was perturbed by an equivalent adjustment of *A* and *B* only, which is a comparable perturbation of the original equilibrium produced by adding *x* to *A*.

There is a relatively small increase in species *A*. because Δ is negative. and a small decrease in species *B*. Although the small decrease in species *B* at first appears counterintuitive, these are the overall changes that would perturb the original equilibrium without an addition of *x*, which by the conservation of mass law, requires a concomitant reduction in species *B* if species *A* is increased.

As Δ returns to zero, the system returns to the original equilibrium condition by an adjustment, *y*:

$$Keq = \frac{[B]}{[A]} = \frac{[B + y]}{[A + x - y]}$$

And solving for *y* gives:

$$y = \frac{[B]x}{[A] + [B]}$$

which will restore the original equilibrium.

NOTE

The above paragraphs are extracted, with modifications, from R. G. Lanzara, "The Quantification of Le Chatelier's Principle," *Theoreticalpharmacology*, blog, February 2, 2018, https://theoreticalpharmacology.wordpress.com/2018/02/02/the-quantification-of-le-chateliers-principle/.

APPENDIX III C: SENSES AND RECEPTORS

THE BALANCE: BALANCING SENSORY AND CHEMICAL TENSIONS

B eginning with a simple, two-pan balance, the side of the balance with the heavier weight will tip toward that side of the balance. When the weights on each side are equal, the balance is in the horizontal position. This concept has been used throughout antiquity to measure and trade various goods. However, what has gone unrecognized is that if the balance is tipped to the maximum so that one pan rests on the table top, and if we then add heavier weights *equally to both sides of the balance*, the balance's maximally tipped pan will rise off the table top.

Although this must have been noticed before, there were no noticeable efforts to explain this behavior in scientific or physical terms. Ironically, I found an equation that explained the behavior of the balance by describing the two possible on and off states of signaling molecules acting on their target receptor molecules. In pharmacology, we study how these molecules transmit signals to their target receptors. These signals include light, pressure, and other molecules, called ligands, which include hormones and some drugs that bind to these receptor molecules.

After years of wondering why this worked, I discovered that it is, essentially, a mathematical identity that also works for the net shift of a simple, two-pan balance. That's when I discovered and wrote a paper on how a well-known physiological law, Weber's law or the Weber-Fechner law, is obeyed by both our sensory receptors and a simple balance. This provided an important link between the physical, chemical, and biological realms that are being intensely studied today.

This simple balance model describes many classical dose-response curves in pharmacology, as well as the phenomena known as spare receptors and rapid desensitization. This is also the basis for the negative feedback mechanism, which, as mentioned before, is vitally important for life. These observations of the simple balance are important in understanding how life began from chemistry. They provide perhaps the first and most primitive mechanism for negative feedback control. Even the simple balance can display negative feedback control!

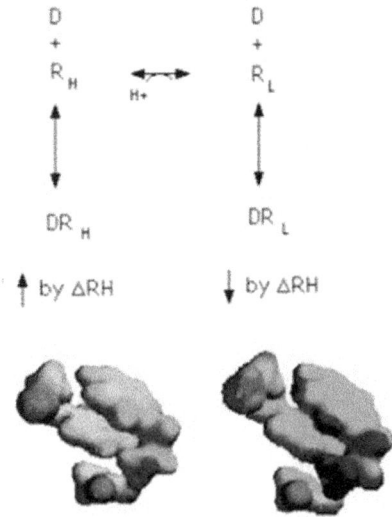

This chemical principle basically states that if we change one side of a chemical equilibrium, then the other side will produce a net shift in the underlying equilibrium that compensates for that change. This net shift is a change in the relative proportion of the underlying equilibrium states. This represents a new type of chemistry based upon the parameterization of Le Chatelier's principle. As shown in Fig. 1, "The Chemical and Molecular Origin of the Net Shift," to the left, this net shift can be calculated in a unique way with the parameter Δ or ΔRH. I subsequently found that by using simple binding expressions for how a molecule binds to each of its target-receptor's two states and equating this to a ratio that accounts for the initial transfer of one state to another, I could measure the pharmacological responses as the net shift in these systems. This shift in the underlying equilibrium was an important Essential Tension that must have been necessary for the formation of all life because it provides life with a relatively simple mechanism to react and respond to changes in the environment.

These shifts largely comprise how our senses respond and, ultimately, adapt to environmental changes. To quantify this equilibrium shift (net shift) in chemical systems, I've provided a strictly mathematical derivation below.

THE MATHEMATICAL DERIVATION

Given any mathematical ratio: $\dfrac{a}{b}$, the expression: $\dfrac{a}{b} = \dfrac{a}{b}$, can be changed to the following mathematical identity:

$$\frac{a+x}{b+y} = \frac{a+\Delta}{b-\Delta}$$

Where x and y represent any values, and Δ represents the conservation of mass relation as the transfer of some amount from the denominator b to the numerator a. Solving for Δ gives:

$$\Delta = \frac{bx - ay}{b + y + a + x} \qquad (1)$$

Where Δ represents a specific addition to a and a corresponding subtraction from b. This is a **mathematical identity**—that is, any fraction can be changed by adding any numbers to the numerator and the denominator or by equivalently subtracting an amount, Δ, from the denominator and adding it to the numerator. This is a fundamental equation that also calculates how much an equilibrium is altered (net shift) from the original equilibrium upon the additions of x and y respectively to a and b.

These relatively simple physical experiments demonstrate something very profound about a two-pan balance that to my knowledge hasn't been adequately explored or addressed before. This discovery points to a fundamental equation of equilibrium that may have been previously discovered but was lost. Digesting this requires some time, but those who try will be rewarded by the elegant simplicity of this fundamental equation that may hold the key to several complex physical, chemical, and biological problems.

NOTE

The equilibrium shift has been previously noticed as a separate area of chemistry, but it was not parameterized. See Peter T. Corbett et al., "Dynamic Combinatorial Chemistry," *Chemical Reviews* 106, no. 9 (2006): 3652–3711, doi:10.1021/cr020452p, https://core.ac.uk/download/pdf/12925202.pdf. The history of this concept is also presented in the above-referenced article.

REFERENCES

Bhat, V. T. et al. "Nucleophilic Catalysis of Acylhydrazone Equilibration for Protein-Directed Dynamic Covalent Chemistry," *Nature Chemistry* 2, no. 6: 490–497, doi:10.1038/nchem.658, https://www.research.ed.ac.uk/portal/files/8667214/ukmss_31507.pdf.

Hioki, Hideaki, and W. Clark Still. "Chemical Evolution: A Model System That Selects and Amplifies a Receptor for the Tripeptide (D)Pro(L)Val(D)Val," *Journal of Organic Chemistry* 63, no. 4 (January 22, 1998): 904–905, https://pubs.acs.org/doi/abs/10.1021/jo971782q.

Lanzara, Richard G., "Method for Determining Drug Compositions to Prevent Desensitization of Cellular Receptors," US Patent, US5597699A, 1992-09-301997-01-28.

——. "Weber's Law Modeled by the Mathematical Description of a Beam Balance." *Mathematical Biosciences* 122, no. 1 (July 1994): 89–94, https://doi.org/10.1016/0025-5564(94)90083-3 and http://cogprints.org/4094/1/Weber's_Law.pdf.

——. "A Novel Biophysical Model for Receptor Activation," *Canadian Journal of Physiology and Pharmacology* 72, supplement 1 (1994): 559 (Abstract P18.13.10).

A beam balance obeys Weber's law and also displays desensitization, which are both common characteristics of sensory and receptor systems. Therefore, a model of the chemical balance between two receptor states was explored by analogy to a physical beam balance. By this model the description of a net chemical torque offers a novel biophysical method to describe how ligand binding to two-receptor states produces a response. The net chemical torque or shift is analogous to the shift produced by weights applied to a physical balance but results from the relative difference between the affinities of a ligand for the two receptor states. An antagonist-ligand has similar affinities for the two receptor states and therefore causes no shift, whereas an agonist-ligand has different affinities for the two receptor states. An equation for the net amount of receptor activation was developed by combining the ligand-binding equations for each receptor state independently with the equation for the shift of a balance. This balance model produces biophysical descriptions of ligand efficacy, rapid desensitization, pH-dependent response and spare receptors. Experimental dose-response curves from Traynelis and Cull-Candy; [*J. Physiol.* 443: 727-763 (1991); del Castillo and Katz [*Proc. Roy. Soc. Lond.* 146: 369-381 (1957); Keen and Krane [*Trends Pharmacol. Sci.* 12: 371-374 (1991)]; and Millar and Stephenson [*Br. J. Pharmacol.* 11: 379] are modeled with reasonable biophysical parameters by this approach. In addition, a biophysical mechanism can be presented for receptor coupling with guanine nucleotide-binding proteins (G proteins) or other transducer molecules.

This model was also presented as a poster at the conference: "Receptor Activation by Antigens, Cytokines, Hormones and Growth Factors," The New York Academy of Sciences, October 21–25, 1994, Orlando, Florida.

A QUESTION OF BALANCE? NONLINEAR COMPLEX PHENOMENA IN BIOLOGY AND PHYSICS

In 1834, the physiologist E. H. Weber (17951878) studied the responses of humans to physical stimuli. He discovered that at least a 5% difference in weight was required for people to guess the difference between unequal weights. He hid the weights with a lightweight paper so that the subjects could not see them. If the weight placed in the subject's hands was 100 grams for each hand, then he had to add five extra grams to one hand for people to sense that one hand held the larger weight. However, if the weight was eighty or sixty grams, he had to add four or three grams, respectively, for people to tell the difference. This law, which is also named the Weber-Fechner law, gained wide recognition when it was discovered that many of our sensory perceptions follow it. However, the underlying basis for this law hasn't been clearly understood. Could it possibly be a basic physical law?

If we examine more closely the various ways that a balance can be tilted, then the physical basis for Weber's law may become evident. At the top of Fig. 2 below is an equal arm balance with equal sets of weights in horizontal equilibrium. Shown on the left side of Fig. 2 is one way to tilt this system by placing unequal weights on the right and left pans together with the original weights. This tips the balance toward the side having the most weight and creates an angle α from the horizontal equilibrium. There is an alternative but equivalent way to produce angle α, which is by moving some of the original weight from one side and placing it on the opposite side as shown in the right half of Fig. 2.

Figure 2. Equivalent Ways to Tilt a Balance to Create Identical Angles α

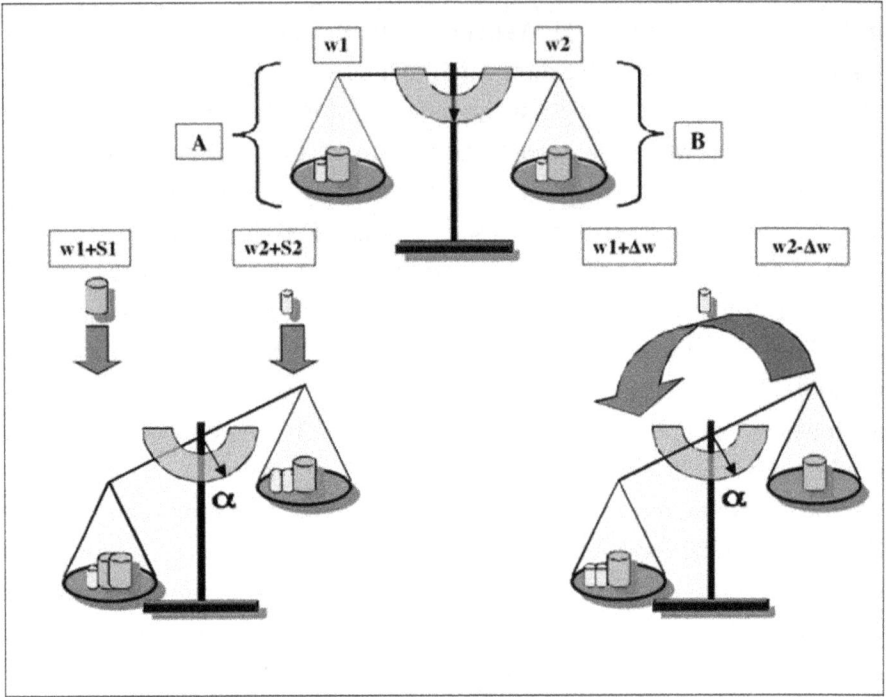

Therefore, we have for an equal arm balance the following equivalent ratios that produce identical angles α:

$$\frac{w1+S1}{w2+S2} = \frac{w1+\Delta w}{w2-\Delta w} \qquad (1)$$

These ratios show why the pan of the balance was lifted off the table by the addition of equal weights. If $w1$ and $w2$ are equally increased, then the ratios will be decreased along with the corresponding angle α. Solving for the transfer of the fraction of weight, Δw, gives,

$$\Delta w = \frac{S1w2 - S2w1}{w1+S1+w2+S2} \qquad (2)$$

where $w1$ and $w2$ are the initial weights in horizontal equilibrium. $S1$ and $S2$ are the additional weights added to each side, as shown in Fig. 2. Equation 2 is a fundamental equation of that equilibrium that measures the net amount of stress applied to the initial equilibrium.

Equation 2 was shown to obey Weber's law (1). Surprisingly, biological receptors compress the sensory functions by a ratio-preserving process that is strictly compatible with Equation 2. At that time, it was also suggested that a modified version of this equation could model the responses of biological receptors.

There is always the impetus to take a simple system and elaborate on it. Therefore, substituting mathematical functions, such as $f(S)$ and $g(S)$, for the parameters S_1 and S_2 in Equation 2 gives,

$$\Delta w = \frac{f(S)w2 - g(S)w1}{w1+f(S)+w2+g(S)} \tag{3}$$

This general expression compares the relative effects of the two functions $f(S)$ and $g(S)$ on an equilibrium system, which allows us to consider more complex variations of Equation 2. Two of these variations are presented below.

More than half a century ago, Langmuir (1881–1957) proposed the chemical binding isotherm equation, as $SR = R(S)/(S+K)$, to be used as a description for the absorption of molecules onto surfaces. Since then it has been used universally in pharmacology and chemistry to describe independent, single-site binding of one molecule to another. If the weights are applied to the pans of the balance, according to the Langmuir equation, then we can measure the stress produced by unequal weighting to the two pans of a physical balance, which is analogous to the unequal binding of a molecule to either side of a chemical equilibrium.

Figure 3. A Two-State Chemical Equilibrium with Binding of Molecule S to the Two States R_1 and R_2.

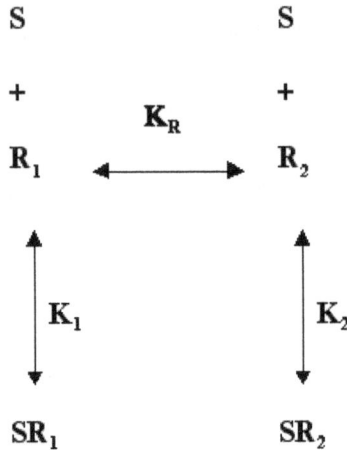

The analogy between the physical and chemical balances requires a more detailed consideration to relate each part of the two systems to one another. As shown in Fig. 3 above, the equilibrium constant, KR, sets the initial amounts of R_1 and R_2. The binding of S to R_1 and R_2 forms SR_1 and SR_2, which will stress the initial equilibrium if K_1 and K_2 are unequal. Linking the physical parameters of the balance to the chemical parameters from the figure, $w_1 = R_1$ and $w_2 = R_2$ and substituting $f(S) = SR_1 = R_1(S)/(S+K_1)$, and $g(S) = SR_2 = R_2(S)/(S+K_2)$ into Equation 3, where K_1 and K_2 are the dissociation constants of the molecule S for R_1 and R_2. Then letting $\Delta w = \Delta R$ yields,

$$\Delta R = \frac{R_1 R_2 (S)(K_2 - K_1)}{R_1(2S + K_1)(S + K_2) + R_2(S + K_1)(2S + K_2)} \qquad (4)$$

Where ΔR represents the change in the amount of "weight" equivalent to the perturbation produced by asymmetrical molecular binding ($K_1 \neq K_2$). This provides a convenient method to calculate the initial stress applied to a two-state equilibrium in terms of competing dissociation constants K_1 and K_2.

When a ligand binds with a greater affinity to one side of a two-state chemical equilibrium, this stresses the initial equilibrium toward the side with the higher affinity. However, this greatly depends upon how we define the chemical species that comprise the chemical equilibrium. This binding preference also produces the phenomenum known as Le Chatelier's principle. However, there is a critical difference between Le Chatelier's principle and. Le Chatelier's principle states that the original equilibrium will shift to relieve the stress applied to the equilibrium, whereas, determines the amount of state that must be transferred to produce an equivalent stress on the original equilibrium.

Equation 4 was tested to see if it could generate responses compatible with those for biological receptors. For this demonstration, (S) represents an amount of weight available for Langmuir binding to R_1 and R_2, which is similar to the idea that the chemical concentration represents an amount of a chemical species available to combine with another chemical species. The dissociation constants, K_1 and K_2, are arbitrarily set to 10 and 100 (all units are in grams), which represent the unequal binding affinities ($1/K_1$ and $1/K_2$) of (S) for pans A and B. The numbers aren't important, they are easy to adjust for specific examples and are provided here for demonstration purposes only. R_1 and R_2 were each set equal to 100, and the amount of (S) was allowed to vary up to 500.

Figure 4. Plots of from Equation 4 also Showing Plots of the Weight on Each Pan from Langmuir Binding (Pan A and Pan B) and the Total Weight (Pan A+B).

(A)

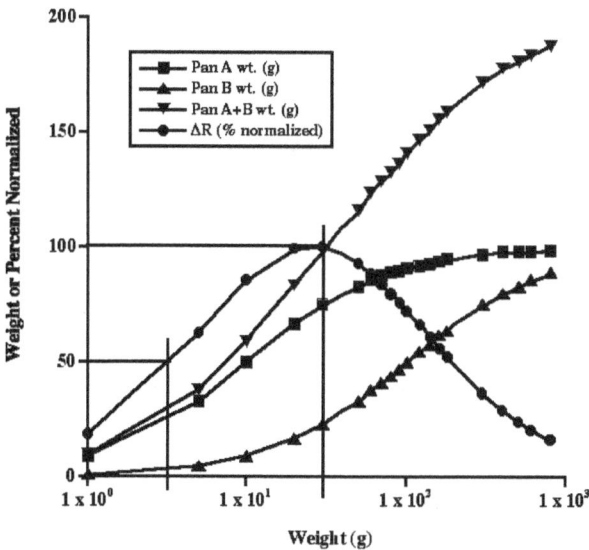

(B)

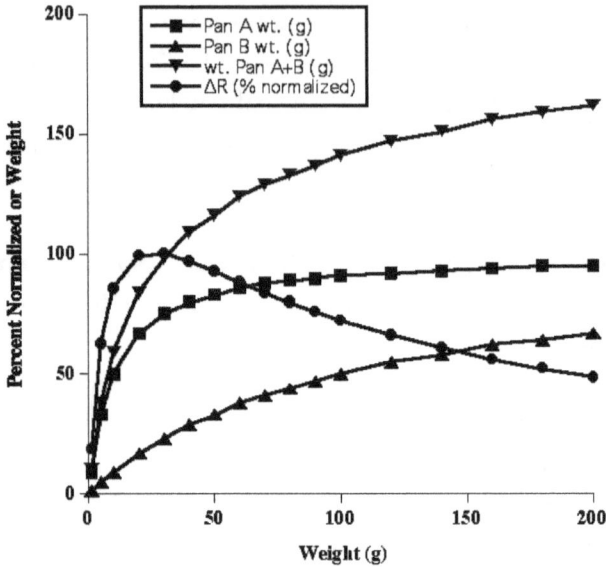

Note: (A) = logarithmic x-axis; (B) = linear x-axis

These plots in Fig. 4 show the nonlinearities in the plots from Equation 4. In Fig. 4(A) and (B), the curves for Pans A and B show hyperbolic binding as expected for Langmuir binding curves. These plots are characteristic of binding curves seen in several biological and pharmacological experiments. The plot for on the logarithmic scale shows a bell-shaped curve that rises to a maximum and declines. On the linear scale (Fig. 4 (B)), the plot of shows a curve that rises to a maximum and then gradually declines. These are common, nonlinear patterns seen in many experiments that measure the responses of biological systems.

Figure 4 (A) and (B) also displays several other characteristics that are unique to response curves. First, the maximum of is below the maximum values for any of the other curves. In Fig. 4 (A) the straight lines indicate the positions on the curve where the 50% and 100% responses occur. The 50% response for occurs at about three grams and the 100% response occurs at about thirty grams. Since these points occur where there is a relatively small fraction of the total binding, this suggests a physical rationale for the phenomena of spare receptors, which is a phenomenon in pharmacology that has puzzled pharmacologists for decades.

Second, the curve for declines with the addition of extra weight. This shows the surprising finding that a physical balance desensitizes. Desensitization, which is the fade of the response in the presence of continuous stimulation, is an essential physiological mechanism that regulates our responses to hormones and appears in a number of important biological receptors.

There is also evidence at the molecular level that the two chemical states R_1 and R_2 result from the pH-dependence of a common, pH-dependent residue within receptors. (2). We have previously constructed a two-state molecular model showing these two chemical states as acid and base equilibrium. Fig. 5, below, shows the molecular electrostatic potentials of these states along with a potential binding molecule. In this molecular model the acid and base states act as the switch for receptor activation (Fig. 5). Agonist ligands activate receptors by showing a preference for the base state. This is the more electronegative state shown in red in Fig. 5 that attracts the positively charged end of the binding molecule shown in blue. This preferential attraction of the binding molecule for the base state places a stress on the original receptor equilibrium that is registered at the receptor either by a shift of the equilibrium or by a change in the underlying dynamics of the receptor.

Figure 5. Acid and Base States of the Molecular Model for the Two-State Chemical Equilibrium for Receptor Activation.

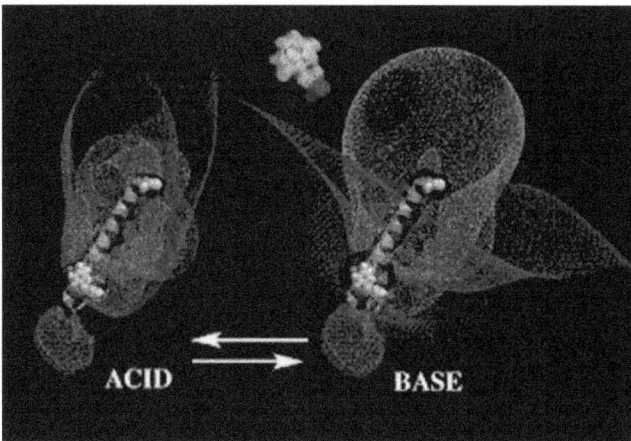

The molecular electrostatic potentials are plotted as positive 25 (blue) and negative -25 kJ mol^{-1} (red) meshes. A potential binding molecule (multicolored) is also shown approaching the acid and base states.

Extending the balance analogy further, the phenomena of inhibition and inverse agonism can be tested by adding the factor of $(1+[I]/K_i)$ for a competitive antagonist, [I], binding to each state with the dissociation constant, K_i. For competitive inhibition of Langmuir binding, this factor is multiplied times each of the dissociation constants, K_1 and K_2. The Langmuir functions then become, $SR_1 = R_1(S)/(S+K_1(1+[I]/K_i))$, and $SR_2 = R_2(S)/(S+K_2(1+[I]/K_i))$. An example of this is shown in Fig. 6, below, for " + [I](-7,-9)" and " + 10[I](-7,-9)", where " " is shorthand for . The two plots of with the inhibitor, [I] and 10[I], display the parallel shift to the right typically seen in competitively inhibited dose-response curves. However, there are often examples where the inhibitor does not have exactly equal affinities for each state. If we allow the inhibitor to have different dissociation constants in place of the single K_i, we can create series of plots that show a wide range of variations seen in many biological and pharmacological dose-response experiments (Fig. 6).

Figure 6. Plots of, Replaced by from Equation 4 with the Addition of the Expressions for the Competitive Inhibitor, [I].

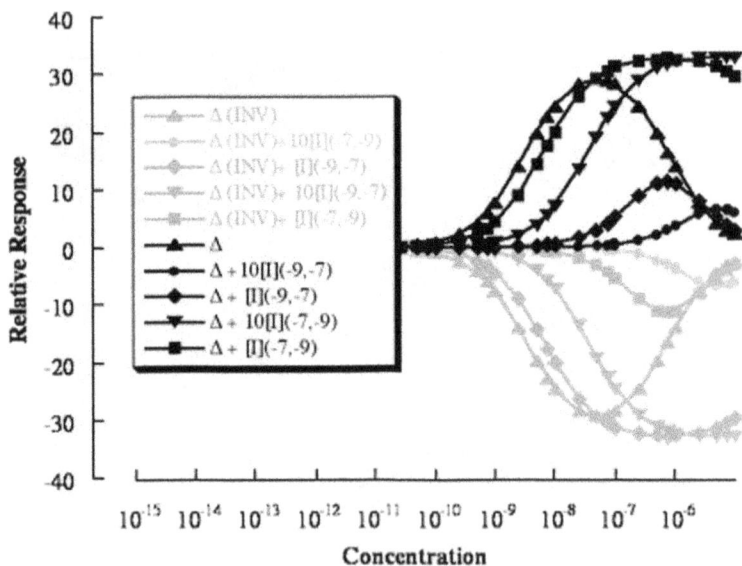

In the legend, the numbers in the parentheses after [I] replace the single parameter K_i. These numbers represent the exponential values for the dissociation constants of the inhibitor for each receptor state, R_1 and R_2. The (INV) plots, shown in gray from Equation 4 with $K_1 \geq K_2$, which makes negative.

If the K_1 and K_2 values are reversed, then Equation 4 is negative, which produces the inverse agonist responses as shown in gray in Fig. 6. This suggests that the binding ligand prefers the other state, which has been previously observed for inverse agonists. Surprisingly, our simple balance model appears to describe the relatively complex nonlinear properties of inverse agonism and modulation observed in many biological receptors. Also, this model has previously described receptor activation, fast receptor desensitization, and a general method for preventing desensitization.

The overall analogy suggests that we are measuring something fundamental to both a physical balance and the chemical equilibrium of biological receptors. In this respect, the weighting of a balance corresponds to the perturbations of ligand binding to receptors. How the underlying equilibrium in either system becomes perturbed is the core concept most important to measure.

What about testing other competing sets of functions? If instead of using Langmuir functions, we examine the two Gaussian functions,

$$f(S) = SR_1 = R_1 e^{-(S-\mu_1)^2/\sigma_1^2} \text{ and. } g(S) = SR_2 = R_2 e^{-(S-\mu_2)^2/\sigma_2^2} \text{ and}$$

substituting into Equation 3 gives,

$$\Delta R = \frac{R_1 e^{-(S-\mu_1)^2/\sigma_1^2} w_2 - R_2 e^{-(S-\mu_2)^2/\sigma_2^2} w_1}{w_1 + R_1 e^{-(S-\mu_1)^2/\sigma_1^2} + w_2 + R_2 e^{-(S-\mu_2)^2/\sigma_2^2}} \tag{5}$$

For some arbitrary values, the resultant has a positive and negative side as shown in Fig. 7 below. The values of the parameters are not particularly important for this example. The important point is that we can substitute a new function in place of the Langmuir functions and achieve another interesting result. Just as the balance can move either up or down, so the graph of shows that weighing competing Gaussian probabilities produces a sine-like wave.

Figure 7. Plots of the Gaussian Functions and from Equation 5.

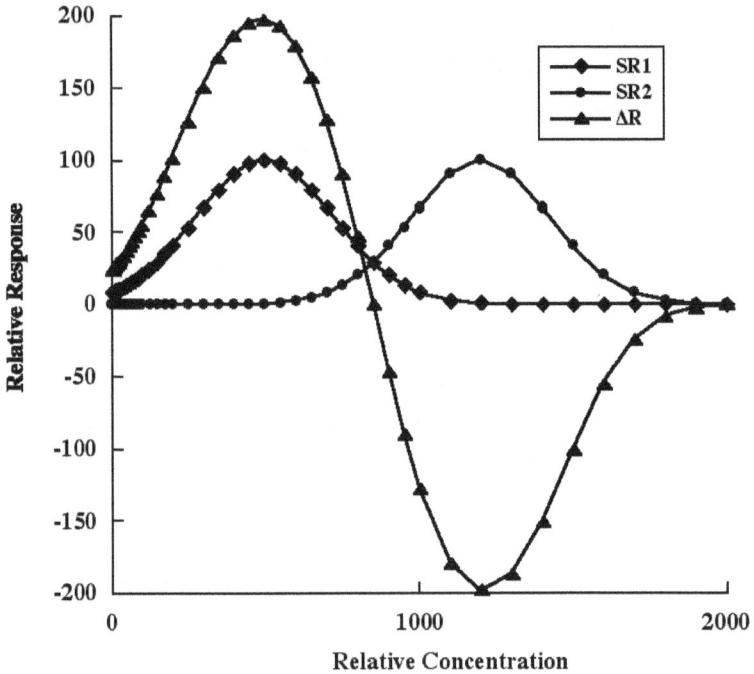

The x-axis is arbitrarily labeled "Relative Concentration," but this could have just as easily been labeled "Relative Probabilities," depending on the interpretation given to the functions of S. It is interesting to consider that two competing Gaussian probabilities yield a new character, , that can be negative and describes the stress placed upon the underlying equilibrium of the probabilities between the two states.

Some of these examples are more developed than others, but the important point is that a fundamental equation of equilibrium derived from a simple balance provides new insights into complex phenomena in biology and physics. The extension of this approach to other areas may prove fruitful as well.

NOTE
The above material is a modified extract from Richard G. Lanzara, "A Question of Balance? Nonlinear Complex Phenomena in Biology and Physics," *Ombamltine*, blog, September 15, 2017, https://ombamltine.blogspot.com/2017/09/a-question-of-balance-nonlinear-complex.html.

REFERENCES
1. Lanzara, R. G. "Weber's Law Modeled by the Mathematical Description of a Beam Balance." *Mathematical Biosciences* 122, no. 1 (August 1994): 89–94, http://cogprints.org/4094/.

Wikipedia, s.v. "Weber–Fechner law," https://en.wikipedia.org/wiki/Weber%E2%80%93Fechner_law.

——. "How Understanding Weber's Law Led to an Understanding of the Receptor Response." LinkedIn, October 5, 2017, https://www.linkedin.com/pulse/how-understanding-webers-law-led-receptor-response-richard-g-.

——. "Optimal Agonist/Antagonist Combinations Maintain Receptor Response by Preventing Rapid Beta-1 Adrenergic Receptor Desensitization." *International Journal of Pharmacology* 1, no. 2 (2005): 122–131. (These were technically difficult studies that require me to acknowledge the important assistance of Drs. Judith K. Gwathmey and Roger J. Hajjar. Without their assistance this work would not have been possible.) http://www.bio-balance.com/ijp.pdf.

——. "A Quixotic Quest to Change Pharmacology." LinkedIn, July 26, 2017, https://www.linkedin.com/pulse/quixotic-quest-change-pharmacology-richard-g-lanzara-mph-ph-d-?trk=portfolio_article_card_title.

2. Rubenstein, Lester A., and Richard G. Lanzara, "Activation of G Protein-Coupled Receptors Entails Cysteine Modulation of Agonist Binding." *Journal of Molecular Structure (Theochem)* 430, nos. 1–3 (1998): 57–71, https://www.

sciencedirect.com/science/article/pii/S0166128098902172 - http://cogprints.
org/4095/1/Cys_Paper.pdf.

Rubenstein, Lester A., Randy J. Zauhar, and Richard G. Lanzara, "Molecular
Dynamics of a Biophysical Model for Beta-2-Adrenergic and G Protein-
Coupled Receptor Activation," *Journal of Molecular Graphics and Modelling* 25,
no. 4 (December 2006): 396–409, https://www.sciencedirect.com/science/ar-
ticle/pii/S1093326306000465?via%3Dihub.or   https://www.ncbi.nlm.nih.gov/
pubmed/16574446.

APPENDIX IV: QUINTESSENCE

These are some of Michael's notes on the origin and magic of growth and evolution.

D arwin's *Origin of Species* transformed our self-image from a species that is absolutely distinct from animals to one that has evolved from them. But in the last few decades research, results in biology have undermined a basic tenet of Darwinism, leaving scientists scrambling to make sense of how evolution really happens. This is because evolution needs a source of genetic novelty and there is no longer a consensus on what that source is.

A theory of growth called Quintessence offers a new explanation for genetic novelty and its role in evolution. Quintessence applies to all life forms and, in the process, revises the story of evolution and revises our self-image. This unified theory is named Quintessence because it is the fifth requirement that all life forms must have to be alive, beyond metabolism, enclosing membrane, reproduction, and growth. Quintessence is about how species capture the novelty of chance encounters, negotiate through opposing pressures, and grow and evolve by internalizing interactions. These processes are inherited from chemistry when life first formed, and they operate at all levels of life from bacteria to plants to animals to humans to societies. Examples of Quintessence in action include how DNA may have first been made when life first formed, the evolution of simple cells to nucleated cells, how multicell organisms began, how genetic novelty directed during the life cycle leads to evolution of species, how one new species was naturally created in months, how a chance meeting can lead to a marriage, and the birth of a business startup, among others.

As species evolve, two trends emerge beyond just the mandate to reproduce. One is a trend toward increasing resources and energy flow in all their forms and abstractions, including food, money, power, tools, building blocks, and information. The other trend is the openness in all species to connect to life around them. Since the experiences of novel chance encounters are unknowable before they happen, sometimes the fateful results of their interactions can be truly magical, be they wondrous, dramatic, loving, dangerous, or just fun.

Quintessence extends our identity from a species that just lives to survive, compete and reproduce, implicit in Darwinism, to a species that also lives for stimulation, connection, and creation. If we accept this new identity, it changes not only how we understand living with each other but also how we understand our relations with all life forms on our planet. In the end, Quintessence helps ground us on how we got here and opens our imagination on where future evolution in biology and culture may take us.

CONDENSED STORY

Darwin requires variation for evolution to go. The problem is that variation by Mendel's rules only allows variation within the limits of the species gene pool, so no new species can evolve. The modern synthesis points to random point mutations of bases as the driver for evolution and the selection pressure of the environment as the direction. The problem is that point mutations almost always produce either neutral or detrimental changes to the phenotype. With three billion bases in the genome and mutations at ten per generation, any beneficial mutations could not happen fast enough to explain the cluster of gene mutations in the human in just the brain, hand, and larynx of the human genome. Eighteen mutations, all clustered, invalidates the random mutations concept. So what is driving the evolution of species?

Before suggesting an alternative, we first need to explore what is the general nature of growth and evolution? One needs to consider the general properties of homeostasis (stability) and change to understand growth and evolution. What are the important properties of novelty and stability that affect growth and evolution? Homeostasis is all about self-regulation to maintain survival with the current set of organism properties. If homeostasis ruled all organisms, nothing would grow or evolve and soon enough that totally homeostatic organism would die from competition. Change driven by the novel outside interactions needs to happen to enable growth. But if the outside drives change too much, then the organisms would die from instability, since change without regulatory control is unstable. So there is a tension between novelty and stability and neither can totally dominate the other but both are in continuous tension with each other.

Any permanent point change to the organism needs both to be allowed by homeostasis and benefit the organism. Benefit is defined here as improving the energetics of metabolism by either efficiency or by increasing the available energy that can be processed by the organism.

Permanent change to homeostasis can occur by either by regulatory change above a threshold that becomes permanent, like a change to the regulatory mechanism itself or by a chance encounter from outside interactions that fits by increasing the energetics of the internal metabolism immediately. The immediate nature of this fit is based on chemical reaction properties and is the most fundamental mechanism of choice, although it is always involuntary.

Once we accept the first fundamental tension it becomes clear that there are many other forms of dynamic tensions that affect both growth and stability.

INTRODUCING A NEW GENERAL THEORY OF CHOICE, GROWTH AND EVOLUTION CALLED QUINTESSENCE.

How can Quintessence offer an alternative to Darwin's view of evolution? To apply these general ideas back to Darwin's dilemma which is gene mutations, there are two sources of gene mutation drivers: one from internal stress measured due to some consistent living situation that eventually modify germ cell DNA. For example, continuous bone stress throughout life would be measured by stress proteins, which would initiate bone tissue to increase bone growth and would also code that change in retroviruses that would distribute through the circulatory system to germ cells and would enter those germs cells and modify the germ cell DNA. The other drive of DNA mutations is from chance encounters such as is needed to create new cell types with brand-new properties. For example, light sensitive rhodopsin from non-light-sensitive opsin. Another example is the layering of additional metabolic pathways to the main core metabolism or the layering of additional ribosomal features with added RNA folds.

How does Quintessence operate across different levels of life?

What are the similarities and difference in how evolution works across the difference levels of life? The major story of evolution is based on considering how change is made relative to different levels of complexity of an organism. There are four types of evolution in the complexity levels of life:

1. evolution from the origin of the cell to the origin of DNA
2. the splitting of cells in asexual reproduction (horizontal gene transfer)
3. evolution by symbiogenesis or joining of codependent cells or species
4. evolution in sexual reproduction species (vertical gene transfer)

In general the more complex the cell type (the larger the cell), the more energy requirements; the more dependence on the conservation of more primitive parts of the cellular infrastructure, such as metabolism and protein types, the less change in new proteins; the more change in morphology of tissue based on intercellular properties, the longer the time from birth to self-sustaining organism, due to the specific layering of structures that recapitulate their evolution so that that specific organism can survive.

How can properties of dynamic essential tensions be applied to different levels of life from the origin of life to human societies?

What can Quintessence say about life's purpose and identity and the origin of choice?

Quintessence is the fifth requirement that all life forms must have to be alive, beyond metabolism, enclosing membrane, reproduction and growth. Quintessence is about how species capture the novelty of chance encounters, negotiate through opposing pressures and grow and evolve by internalizing interactions.

APPENDIX V: CALCULATION OF THE EXTRA MEMBRANE NECESSARY TO COVER TWO DAUGHTER CELLS COMPARED TO THE ORIGINAL MOTHER CELL

In his notes on Quintessence, Michael writes:

> Cell division starts as an internal polarization of chemical gradients. In the cell, midpoint, a radial chemical reaction punctures the cell membrane and starts a zipper head that grows inward. The zipper head grows new cell membrane on either side of it, so new links in the zipper are added: left membrane, zipper head, and right membrane. When the zipper head growth comes together at the center, the beginning of the zipper breaks, and it unravels until two cells separate.

In the case of a dividing cell, the extra membrane required to cover two daughter cells will be compared to the original amount of membrane necessary to cover the mother cell (1). The equations for a spherical cell are, first, for the volume of the mother cell (Vm):

$$Vm = \frac{4\pi R^3}{3}$$

$$\frac{Vm}{4\pi} = \frac{4\pi R^3}{3}$$

Where R is the radius of the mother cell.

Assuming that the total volume of the daughter cells equals the original volume of the mother cell immediately before cell division.

$$Vm = 2Vd$$

$$\frac{4\pi R^3}{3} = \frac{2(4\pi r^3)}{3}$$

Where r is the radius of each daughter cell. Solving for r:

$$r = \frac{R}{\sqrt[3]{2}}$$

Substituting for r in the equation for the total surface area of the two daughter cells:

$$Total(Sd) = \frac{8\pi R^2}{\sqrt[3]{4}}$$

This represents the increase in the surface area (membrane) necessary to minimally cover the two daughter cells. Comparing this to the original surface area of the mother cell:

$$\frac{Total(Sd)}{Sm} = \frac{\frac{8\pi R^2}{\sqrt[3]{4}}}{4\pi R^2} = \frac{2}{\sqrt[3]{4}}$$

Calculating this gives 1.25992, or in terms of the percent increase in surface are about 126% necessary so that each daughter cell is covered by an external membrane.

For repeated divisions, such that occur when an egg cell is fertilized,

$$\frac{Total(Sds)}{Sm} = \frac{\frac{8\pi R^2}{\sqrt[3]{4}}}{4\pi R^2} = \frac{2^n}{\sqrt[3]{2^{2n}}}$$

This can be reduced to:

$$\frac{Total(Sds)}{Sm} = \sqrt[3]{2^n}$$

For two divisions ($n = 2$) this is 1.5874 or about 159% of the original surface area. For n = 10, this is 10.0793 or about 1008% of the original surface area.

This demonstrates the critical need for additional membrane production to cover the daughter cells in a rapidly dividing embryo. The increase in membrane production might contribute to the force producing the cleavage furrow of the dividing mother cell (2,3).

Interestingly, since the new membrane formed is about 26% between the two daughter cells, this translates to about 13% for each cell, but since 126%/2 = 63%, we should normalize this so that 13/63 = 0.2063 or about 21% of the daughter cell's membrane will be new membrane. This means that about 79% (50/63 = 0.7936) will be part of the older membrane from the mother cell. This may represent an important but overlooked mechanism for a type of nongenetic inheritance. It may also be a potential driving force for unrestrained cell division.

REFERENCES

1. Lanzara, R. "Cell Membrane Material (Evidence for an Increase in Membrane Production as a Driving Force for Cell Division)." *Chemical and Engineering News* 55, no. 47 (November 21, 1977): 72, https://pubs.acs.org/doi/pdf/10.1021/cen-v055n047.p004.

2. Mercier, Romain, Yoshikazu Kawai, and Jeff Errington. "Excess Membrane Synthesis Drives a Primitive Mode of Cell Proliferation." *Cell* 152, no. 5 (February 28, 2013): 997–1007, doi: https://doi.org/10.1016/j.cell.2013.01.043.

3. Osawa, Masaki, and Harold P. Erickson. "Turgor Pressure and Possible Constriction Mechanisms in Bacterial Division." *Frontiers in Microbiology* 9, no. 111 (January 13, 2018): 1–7, doi:10.3389/fmicb.2018.00111.

About the Authors

Michael Kuperstein (1954–2018) pursued his career in both science and business. On the business side he was the cofounder of MobiFlex, an online service that quickly and cheaply created mobile apps. Before that, he founded Metaphor Solutions, a speech recognition service company. In 1994 Michael founded eTrue, a facial-recognition company. In 1988 he cofounded Captiva Software Corporation, the global leader in information-capture solutions.

On the science side of his career, in 1987, Michael invented the world's first neural robot that learns from its own experience. He created five patents in various pattern-recognition technologies, published sixteen journal articles, was an associate editor of the journal *Neural Networks*, and co-authored a book on the neural mechanism of visual orientation. Michael possessed a PhD in Neuroscience from MIT and worked as a postdoctoral graduate in Wellesley College's Biology Department, with Professor Howard Eichenbaum, on the physiology of spatial cognition.

Richard G. Lanzara has pursued his career in academia, science, and business. On the business side, he founded three biotechnology companies (Bio Balance, Efficacy, and Enhanced Pharmaceuticals). Before that, he was a tenured full professor in the Department of Allied Health Sciences at CUNY for twenty years until 2000.

On the science side of his career, in 1992, Richard invented the world's first biophysical method to reduce receptor desensitization using agonist/antagonist combinations in a predetermined calculated ratio of agonist to antagonist. He has four patents in various methods to calculate receptor responses and methods that modulate these responses via a bio-engineered process. Richard has a PhD in Biomedical Science/Pharmacology from Mount Sinai Medical School of Medicine as well as an MPH in Environmental Cellular Chemistry from the University of Michigan School of Public Health.

Index

Habituation, 34, 51, 73, 127-128, 147
Hawking, Stephen, 5
Health, 58, 74, 84, 114, 176
Homeostasis, 11, 21, 32, 34, 50-51, 59,
  62, 71-72, 86, 117, 119, 125, 128,
  135, 146, 168-169
Horizontal gene transfer, 107, 109,
  125, 170
Hydrothermal vent, 10, 19, 30, 65, 77,
  136

IBM, 42
Information, 4, 10, 12, 21, 33-35, 51,
  72, 80, 87, 91, 93-98, 102, 119-120,
  125, 131, 167, 175
Inheritance, 9, 11-12, 18, 55, 60, 75,
  87, 92, 94, 97-98, 101, 108, 116, 139,
  141, 173

Lamarck, 19, 24, 60, 105, 116
Langmuir, 145, 157, 159-160, 162-163
Le Chatelier, 146-149, 151, 159
Luca, 9
Lucy, 9
Lyell, Charles, 22, 24

Mandelbrot, Bernoit, 110-111
Membrane, 4, 11-12, 17, 19-21, 27-35,
  52-55, 57, 61, 64-65, 67-71, 75,
  86-88, 91-93, 97-102, 111, 118-119,
  121, 125, 127, 128-129, 134, 136,
  138, 145, 167, 170-173
Metabolism, 12, 15, 18, 20-21, 27-29,
  32-33, 40-41, 47, 50, 52, 54-58,

61-62, 65-67, 71-73, 76-77, 80,
  85-89, 92-93, 102-104, 107, 114,
  118-119, 125, 134-136, 138-139,
  167, 169-170
Mitochondria, 31, 61, 75, 78, 97-99,
  103-105, 115
Modern synthesis, 46, 108, 168
Modulation, 72-73, 96, 114, 163, 165
Mutation, 46, 108-109, 116, 168-169

Natural selection, 14, 46, 108-109
Neanderthals, 108
Neosis, 21, 49, 56, 123, 135
Net shift, 11, 34-35, 51, 72, 96, 129,
  145, 148-152
Neuroscience, 3, 7-8, 175
Nonlinear, 30, 35, 91, 111, 145, 155,
  160, 163, 165
Novelty, 11-13, 18, 20, 22, 49-54,
  56-58, 60, 62, 88, 91, 97, 102-104,
  109, 118-121, 123-125, 135, 137,
  139, 167-168, 170
Nutrients, 20, 66, 83, 89-90, 135
Nuclei, 97, 117
Occam, 94
Odorant, 95
Organelles, 78, 115
Oscillation, 30, 73
Oxygen, 33, 50, 68-69, 73-74, 83-85,
  117, 123, 142-143

Pasteur, Louis, 16
Pharmacology, 7, 34, 146, 149-150,
  153, 157, 160, 165-166, 176